Sailpower

Sailpower

THE SCIENCE OF SPEED

Lawrie Smith & Andrew Preece

fernhurst
BOOKS

www.fernhurstbooks.co.uk

First published in 1994 by
Fernhurst Books, Duke's Path, High Street, Arundel, West Sussex BN18 9AJ

Printed and bound in China through World Print

British Library Cataloguing in Publication Data
A catalogue record for this book is available from the British Library.

ISBN 0-906754-99-2

Acknowledgements
The author and publishers would particularly like to thank the following people who contributed to this book: Peter Bentley, Tim Fletcher, Alan Gray, Carey Mortimer, Ed Dubois, David Howlett and the crew of *Jamarella*.

Photographic credits
All photographs by John Woodward except for the following:
Julia Claxton: pages 15, 34, 60, 77.
Tim Hore: pages 33, 62, 63, 85, 107.
PPL: page 11.
Yachting Photographics: pages 37, 104, 125.

DTP by John Woodward
Artwork by Jenny Searle
Cover design by Simon Balley

Contents

Introduction

America's Cup winner Bill Koch once said that yacht racing was 90 per cent science and 10 per cent art. He later, out of respect for a crew who had just won him the 1992 cup, modified that statement to give them due credit – but he still rated science higher than 50 per cent.

Some may dispute his figures, but the principle of his statement is valid. Yacht racing is largely a science, of a complexity that, to those of us who try to understand it, seems comparable with the science of spaceflight. It is a science of aero- and hydro-dynamics and a science of structures and mechanics. Given the vagaries of the wind and the waves, it is also a science that is still far from exact. Yet to the sailor, even the most intuitive sailor, the science of speed is fundamental to success in yacht racing.

In this book we have tried to describe and explain the technical elements of yacht tuning and racing, and we hope that this will give you a broad understanding of the principles that can be applied in any given situation. We have obviously made simplifications: even the most technical treatises on the complex subject of naval architecture make simplifications and assumptions. We make no apology for this because, in so doing, we hope to have extracted and distilled the elements and principles that are essential to an understanding of the subject – an understanding that is essential if you are to race faster.

For only when you understand the science – and are able to apply it in whatever situation you find yourself – will you be able to concentrate fully on the art.

1 What makes a fast boat?

What is a sailboat?

In technical terms a sailboat is a mechanism that operates on the interface of two media – usually and hopefully air and water – and uses the relative movement between the two media to create a driving force and hence motion.

A sailboat is also a structural unit which must float and must be capable of withstanding the loads imposed upon it by its driving force components – the rig, keel and rudder – as well as the forces imparted by the waves. The unit must also offer the minimum possible resistance to forward motion in as many different conditions as possible.

The basic motion of any sailboat is downwind: it is simply blown in the direction of the wind. However, by using sophisticated methods the boat can be made to move at right angles to the wind and even to some extent towards it. This is achieved by opposing the forces created by the action of the wind over the sails with a force created by the keel or centreboard.

Imagine a lemon pip squeezed between your thumb and forefinger. Your thumb is the keel and your finger the sails. The speed at which the pip is squeezed out depends upon several factors:

1 The angle between your thumb and forefinger, which represents the angle between the wind and the boat.

2 The slipperiness of your fingers and the pip, and its actual shape; these represent the hull shape, its surface finish and rig efficiency.

3 The force at which your fingers are squeezed together, which represents the windspeed.

▼ *A is the resultant force created by the air flowing over the sail. This is split into B, the side component (mostly resisted by the keel), and C, the useful driving component.*

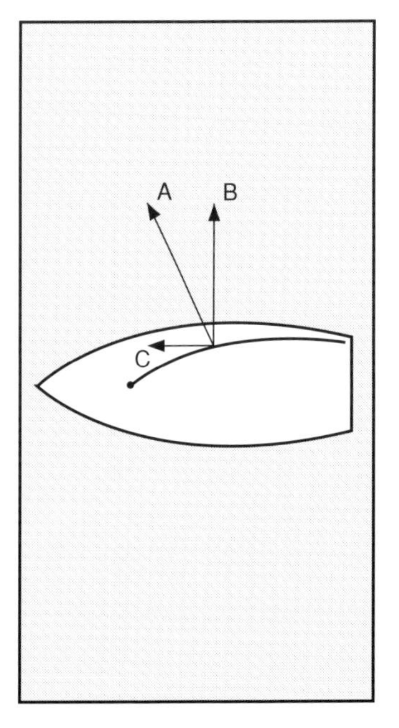

What are the different types of sailboat?

There are two basic types of sailboat in widespread use: monohulls and multihulls. Multihulls can be subdivided into catamarans, trimarans and proas; all of these are outside of the scope of this book.

Monohulls can be divided into keelboats and dinghies. The difference between the two is that a dinghy uses the weight of the crew to provide stability, while a keelboat uses some form of static ballast carried in the keel, or both in the keel and inside the structure of the yacht.

Dinghies and keelboats can be further subdivided into one-designs, where each dimension and parameter is fixed – including, in some classes, which sailmaker may be used – restricted classes where there is some latitude in certain dimensions, and open classes where there are very few restrictions.

Some dinghies and keelboats are designed for particular styles of racing in particular conditions. In England, for example, the Thames A-Rater was designed with a tall, slender sailplan to make best use of the wind passing over the high banks and trees of

London's River Thames. Similarly on the west coast of the United States the maxi 'sleds' are long, light boats with modest sail areas designed for the downwind surf ride from the west coast of America to Hawaii. Both types of boats have evolved to suit their own particular environments and might well perform disappointingly elsewhere.

▲ *The lateral force exerted by the wind is counteracted by the hull and keel.*

What external forces act on a sailboat?

The forces involved are essentially the driving forces created by the wind and the resistance and side forces created by the rig, hull, keel and rudder combination. As the wind passes over the sails a net resultant force is created which can be resolved into two components – one in the direction of motion, and one perpendicular to the direction of motion. The component of the force in the direction of the motion is obviously the component which creates the forward movement.

The other component, at right angles to it, has to be resisted by a sideforce from the keel and hull to create forward motion and prevent the boat slipping sideways. The ratio of the forward component to the sideways component is a measure of the efficiency of the rig and its setting under specific conditions: a correctly-tuned boat sailed well will have a higher forward to sideways (lift to drag) ratio than a poorly-tuned boat sailed badly. The ratio also varies with the angle of sailing, being at a minimum when most of the force is sideways (upwind) and at a maximum when the force is forwards (running).

But the keel must not only resist the sideforce exerted by the sails, it must offer minimum resistance to forward motion as well. Its own force can be resolved into the component opposing the

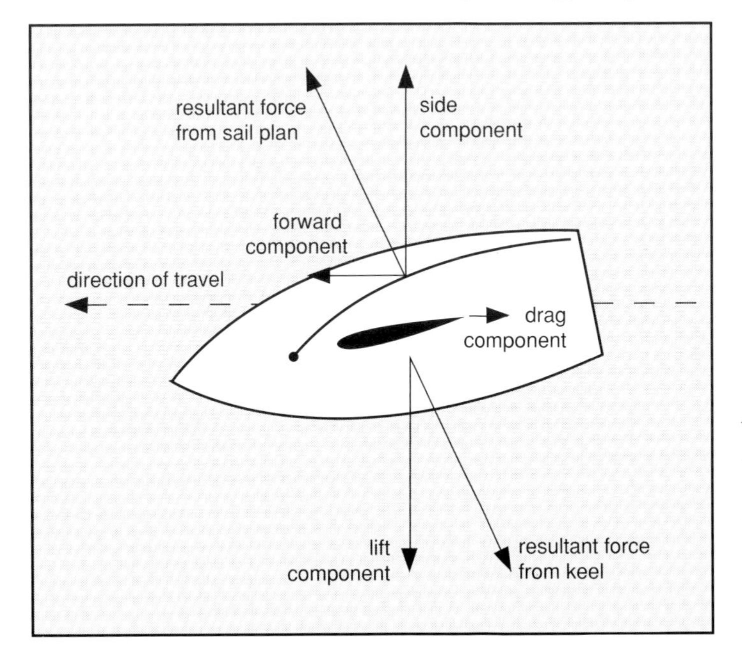

◄ *The symmetrical keel must be set at an angle of attack to develop lift, so the boat inevitably drifts to leeward (makes leeway). The boat accelerates or decelerates until the forward component equals the drag component.*

sail sideforce – the lift force – and its component in opposition to the direction of travel – a combination of several different elements known as the drag force. The drag force of the hull and appendages (keel and rudder) must be added to the air resistance of the hull, deck, fixtures and crew, mast and spars to give the total drag force which, for steady forward motion, must be equalled by the driving force of the sails.

What is the significance of the driving and drag forces?

The entire process of yacht racing and tuning is concerned with minimising the resistance forces and maximising the driving force.

What affects motion through the water?

The speed at which a boat sails through the water depends upon the wind speed, the angle of sailing relative to the wind, the efficiency of the rig, the efficiency of the hull and the effects of any waves the boat might be sailing in. Waves don't necessarily slow a boat down *per se*, but they do affect the flow of wind over the sails and water over the keel, reducing their efficiency. Both the windspeed and the prevailing waves are outside the control of the boat's crew, but the angle to the wind, the efficiency of the rig and the efficiency of the hull are all tunable to a greater or lesser degree.

A sailing boat is a machine of compromise: the best hull and rig configuration upwind in flat water will not necessarily be the best downwind in a rough sea, and similarly a boat that is fast in light winds will not necessarily be fast in strong winds. It is up to the designer to define the conditions for which a boat is particularly required, or to design the best compromise for a range of likely conditions.

Once the hull is built its shape is usually unmodifiable, so the finish of the hull, the tune of the rig and the set of the sails are the only 'gear changes' available. Plus, of course, the style of sailing.

What are the important factors?

Assuming a boat's designer has created an efficient design, it remains for the crew to set that boat up correctly, learn its peculiarities and learn to sail it as quickly as possible. Leading British sailor Eddie Owen reckons he can get a new boat to around 95 per cent of its potential on the first sail, but getting the critical last five per cent might take the rest of a season.

When tuning a boat there are a number of distinct areas which must be considered both individually and together. These are:

- hull weight
- structural integrity
- weight distribution
- surface finish
- foil shape and condition
- rig design and tune
- sail shape.

Hull weight

Hull weight and structural integrity are usually beyond the control of the sailor, yet they have a fundamental effect on performance. In fact, hull weight and structural integrity are usually in conflict because a strong, solidly-built boat will usually be a heavy boat.

In a one-design or restricted class there is usually a minimum hull weight specified by the class rules. It is essential for the boat to be as near as possible to the minimum weight, preferably with the maximum permitted proportion of that weight in the form of correctors which can be removed as the boat ages and gains weight naturally.

The weight of a hull affects the following:

Acceleration For a given driving force the acceleration of any object depends upon its weight, because:

force = mass x acceleration (or acceleration = force/mass)

Therefore, the lower the mass for a given force, the higher the acceleration).

Speed Although the maximum potential displacement speed of a hull is not dependent upon its weight but its length, a heavier hull requires more power to drive it to a specific speed because the hull is displacing and hence 'pushing' more water.

Speed is a curious commodity because it depends upon different factors according to the circumstances. It should be split into two categories – displacement speed and planing speed.

Displacement speed is the speed of a hull when it is immersed in water rather than riding on it (swimming rather than waterski-ing). In the displacement condition the maximum potential speed of a hull is related to the wave system it is creating. A wave system can travel through the water at a finite maximum speed which

▼ *A hull approaching maximum displacement speed through flat water. Note the wave crests at bow and stern and the trough amidships.*

depends upon its wavelength (the distance from crest to crest). At low speeds a yacht may create a wave system with many crests along its waterline length, but as it accelerates the waves grow higher and the crests get further apart until there is one at the bow and one at the stern. At this speed it is impossible for the boat to travel any faster; it has reached its maximum displacement speed.

Planing speed is achieved when a hull rises over its bow wave and travels on top of the water. To achieve this the hull has to generate an upward force – lift – so flat hulls plane more easily than rounded ones. Rounded hulls are usually more efficient in the displacement condition, however, so most hull designs are a compromise in this respect.

In theory there is no upper limit to a hull's potential planing speed, but in practice wind resistance, rig efficiency or engine and propeller efficiency limit the top speed. (The components of overall resistance are discussed further in Chapter 2.)

Speed in waves This is affected by the concentration of weight: a boat that is heavy in the ends, or which has a heavy mast – or indeed keel – will pitch more than a boat with its weight concentrated in the middle. Pitching reduces the efficiency of the sailplan and absorbs energy that would otherwise be converted into driving force. So for this reason alone it is desirable to carry the maximum weight in correctors concentrated around the middle of the boat, and to concentrate weight in the boat where possible.

Structural integrity

The structural integrity of a boat's hull consists of five essential elements:

The ability to float This is a small but important detail!

Longitudinal stiffness The ability of the boat to withstand fore-and-aft rig tension affects the headstay tension and thus performance upwind. For example, modern 40-foot racing yachts carry about 10,000lb of tension in the headstay at maximum. That's the equivalent of more than three VW Golfs hanging off the headstay! Boats such as 470 dinghies have been notorious in the past for going 'soft' very quickly because the hull design was not stiff enough to stand up to the rig tensions necessary for fast sailing at the front of the fleet. This is one reason why new boats are generally faster than old ones.

Athwartships stiffness This enables sufficient athwartships rig tension to be applied to keep the rig in column.

Torsional stiffness The hull has a tendency to twist owing to the fact that its buoyancy, the weight of the crew and the loads of the rig and keel all act at different points along its length.

Panel stiffness This is the stiffness of each individual unsupported area of the hull skin. The greater the panel stiffness the more efficiently the desired hull shape is maintained as the boat moves through the water and impacts with waves. Any distortion of individual panels not only reduces the efficiency of the hull shape but also causes driving force to be absorbed by panel distortion or vibration. It can also lead to fatigue failure of the panel.

◀◀ *The hull starts to plane when it rises over its bow wave. The flatter the hull form the more readily it will plane, but because flat hulls are less efficient at displacement speeds most yacht hulls are compromise designs.*

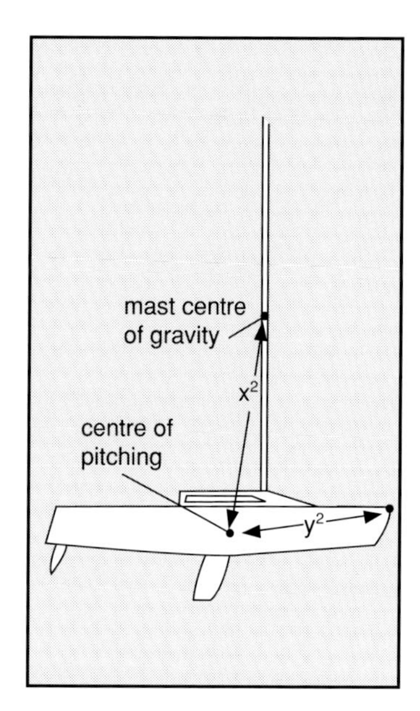

▲ *Weight in the bows (such as an anchor) affects pitching by a factor of y^2. Weight in the mast, acting at its centre of gravity, affects pitching by a factor of x^2. Even weight in the keel has an adverse effect on pitching!*

Although we have considered these structural considerations separately they are all interlinked. The panel stiffness has some effect on the other elements, and the framing required to sustain rig tension has some effect on panel size and strength.

Weight distribution

The speed of a boat in waves is profoundly affected by the distribution of the weight within the boat, and the reason is pitching. Aerofoils and hydrofoils work most efficiently when the flow over them is steady, but when a boat pitches the flow of air over the rig and water over the keel is erratic.

The tendency of the boat to pitch depends on its weight distribution about its centre of pitching. The effect of any item on pitching depends on its weight and the *square* of the distance from the centre of pitching – so a kilogram at the masthead can have the same effect as 50 kg on the deck of an average boat.

So, for the same overall weight, a boat in which the weight is concentrated about the centre of pitching will be faster. Reducing the weight in the ends of the boat, the weight of the rig and the sails, the weight of the keel and the depth of the ballast all help reduce pitching. Model yachts use very deep keels with bulbs at their tips, but model yachts usually sail in relatively flat water. The same configurations on larger craft tend to promote excessive pitching in waves.

Surface finish

There are two significant elements to the surface of a hull: its fairness and its finish. A rippled or unfair surface is slow because it disrupts the steady flow of water around the hull. A poor finish increases the frictional resistance of the hull surface itself.

Over the years there has been much debate concerning the ideal hull surface finish; whether a hull should be as smooth as possible and, if so, what type of paint or finishing material should be used.

It is usually assumed that a mirror-smooth hull finish is ideal, but in the 1987 America's Cup a textured finish known as riblets was tested and found to be very effective. The finish took the form of plastic sheets grooved with microscopic furrows; these were aligned to the flow of the water around the hull and, it was claimed, reduced resistance by aligning the boundary layer which affects the frictional resistance of a hull. The size of the furrows and their alignment with the flow of the water was critical, however, and varies according to hull size and shape. So for all practical purposes they are not feasible, and riblets have been banned in all mainstream branches of the sport.

This leaves us with mirror smoothness, and indeed in the Soling class at the 1992 Olympics it was found that polishing the hull every day made a substantial difference to boatspeed. For boats without antifouling, therefore, polishing is best. Where antifouling is necessary the hull can be smoothed using 600 grade wet and dry abrasive.

Foil shape and condition

Some of the most restrictive class rules leave some latitude for variation in the shapes of rudders and keels or centreboards, both in profile and in section. Some experimentation in section and profile, if allowed, can often improve the lift/drag ratios of the foils.

Like the yacht itself, a rudder or keel design is a compromise which has to perform well in light winds and heavy winds, flat water and rough water, upwind and down, and on both tacks. The nearest a sailing vessel comes to the steady state enjoyed by an aircraft is in speed sailing events when the yacht sails on one tack, at one apparent wind angle and in relatively high winds. The rudders and keels of such boats are tailored to these specific conditions and look quite different from the foils we are used to seeing on cruising and racing boats, being deep and narrow with asymmetric sections to provide optimum efficiency on one tack.

Rig design and tune

Along with the keel and rudder shape, rig design and tune – coupled with sail shape and tune – are the most significant elements over which the sailor has some influence. Like the foils, the rig is a compromise: between strength, weight, section size and the amount of standing rigging required to hold it up, adjust and control it. In simple terms, large-sectioned masts are lighter but create more windage.

It is essential that a mast and mainsail are in harmony, that the sail is cut to the mast and that the mast is capable of being bent to alter the sail for different conditions.

As well as supporting the mainsail, a mast must be able to withstand the high compression loads needed to achieve the forestay tension necessary for upwind sailing.

The chapters on rigs and sails cover all these topics in greater depth.

Sail shape

Sail shape can be divided into two distinct areas: the shape built into a sail and the shape that is controlled and adjusted by the crew. There is obviously some overlap between the two and a crew can sometimes compensate for any deficiencies in the cut by using various sail controls. But although this is possible, a rig which has been contorted to fit a badly-cut sail is unlikely to be versatile enough to be set up properly for a range of different conditions. Therefore it is very useful if a crew can distinguish between a poor sail design and an incorrect set-up, and either discuss a recut with the sailmaker or adjust the rig to suit the sail.

Aerofoils and hydrofoils

Aerofoils and hydrofoils work in essentially the same way, but one – the rig – operates in air while the other – the keel or rudder – operates in water.

Foils convert energy from the flowing stream of air or water into a lift force and a drag force, splitting the steady stream and forcing particles to flow down either side of the foil. Because of the shape of the foil (or its angle to the flow in the case of a symmetrical foil) the particles travelling down the leeward side of a sail – or the windward side of a keel – have further to travel than the particles passing along the other side. The particles that have gone the 'long' way travel faster, and this reduces the pressure on that side of the foil. Consequently there is a higher pressure on the other side, and this creates a lift force.

Although the total force created by a foil can be partially resolved into a lift component, there is also an element which must be resolved into a drag component. The ratio of lift to drag is the subject of continual study and experiment and varies for different foils in different fluids under different conditions.

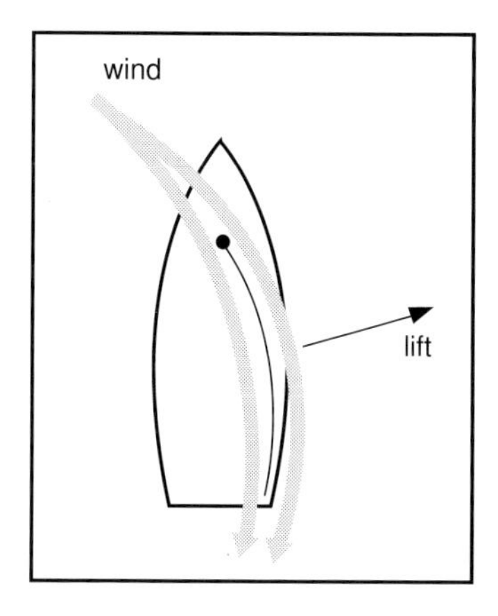

▲ *Air flowing around a sail is separated into two streams, travelling further and faster around the leeward side to create a zone of relatively low pressure and thus a resultant lift force.*

▼ *In the same way, as the yacht makes leeway the keel moves through the water at an angle. The water flowing around the weather side of the foil travels further and faster, creating a low-pressure zone and a resultant lift force.*

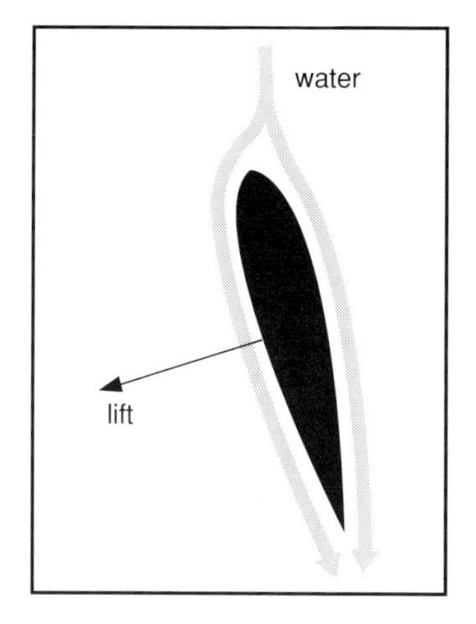

TYPES OF FLOW

The conditions of lift and drag depend on the flow around the foil which can be categorised in three ways: laminar flow, turbulent flow and separated flow. All these types of flow occur in what is known as the boundary layer – the region close to the foil surface in which the particles of water or air are, to varying degrees, being dragged along by the moving boat.

On a keel, for example, there are particles of water actually attached to the foil surface by friction. They are moving along at the speed of the boat, say eight knots, but their relative speed to the boat is zero. Some distance away the particles of water are unaffected by the keel and are effectively stationary, so their speed relative to that of the foil is the same as the speed of the boat – eight knots. The zone between these two extremes is occupied by particles of water moving at relative speeds ranging from zero to eight knots depending on their distance from the keel. This zone is known as the boundary layer.

The particles in the boundary layer can travel past the foil in various ways. If they are displaying laminar flow they move like sheets of paper all sliding over each other around the shape of the foil. In turbulent flow the particles are still adhering to the shape of the foil but they are travelling past erratically. In separated flow the particles are moving extremely erratically, swirling in eddies and not really following the shape of the foil at all.

Laminar flow is the most efficient, since it offers the least resistance. Turbulent flow is less efficient and separated flow is plainly inefficient. Therefore the aim is to increase the ratio of laminar flow to turbulent flow and reduce separated flow to a minimum.

The nature of the flow depends on the viscosity or 'thickness' (as in water versus treacle) of the medium, and its speed. In air, because it is so 'thin', there is no laminar flow past a rig. In water the flow remains laminar for a few inches from the leading edge, but the flow along the remaining length of the foil is turbulent. If the foil were in treacle laminar flow might be maintained further! So while it is desirable to increase the laminar flow along a foil it is difficult to increase it substantially, and certainly any talk of laminar flow keels with the entire keel in laminar flow is rubbish.

▼ *At low speeds the frictional resistance (R_f) is a high proportion of the total. At high speeds the wavemaking resistance (R_W) is the dominating factor. In practice this means that hull wetted area and finish are more significant in light airs and hull shape is more significant in heavier airs.*

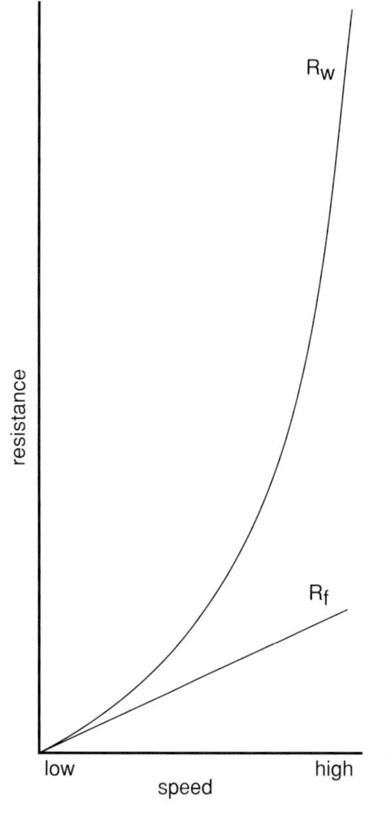

RESISTANCE

Before considering how to make a boat go fast, it is useful to understand what stops it. The culprit is resistance, or drag.

If the boat is to sail at any speed at all the overall drag of the hull, keel, rudder and rig must be equalled by the driving force generated by the rig. However, although the drag can be described as a single force it is actually the sum of a number of different components that fluctuate in significance depending on the conditions.

First the drag can be divided into aerodynamic drag and hydrodynamic drag, as follows: Total resistance = aerodynamic resistance + hydrodynamic resistance ($R_t = R_a + R_{th}$).

The aerodynamic drag is created by the air resistance of the rig, the crew, the deck gear and the hull above the waterline. It is difficult to measure and is more significant at higher windspeeds since it varies with the square of the speed of the wind.

Hydrodynamic drag, or total hull resistance, consists of frictional resistance plus residuary resistance ($R_{th} = R_f + R_r$). The residuary resistance consists of wavemaking resistance plus induced resistance, plus eddy-making resistance ($R_r = R_w + R_i + R_e$).

Firstly there is the friction drag (R_f) of any part of the boat – hull, keel, rudder, propeller – that is immersed in the water. The friction drag depends on the speed of the water flowing past, the surface area of the item in question and the surface finish of the item. So the smaller the surface area and the smoother it is, the lower the drag. At very low speeds, friction drag accounts for a large percentage (85–90 per cent) of the total drag but at high speeds much less (say 30 per cent).

Then there is residuary resistance. The first component of this is wave drag, which occurs at the water surface (fully-immersed keels don't make waves, for instance, but transom-hung rudders do) or at the interface between two media of differing densities (usually water and air). Wave drag is a complex subject which depends on the form of the hull: the weight, length, beam and draft, and the proportions of each relative to the others. Wavemaking drag is less significant than friction drag at very low speeds but becomes the critical factor at higher speeds.

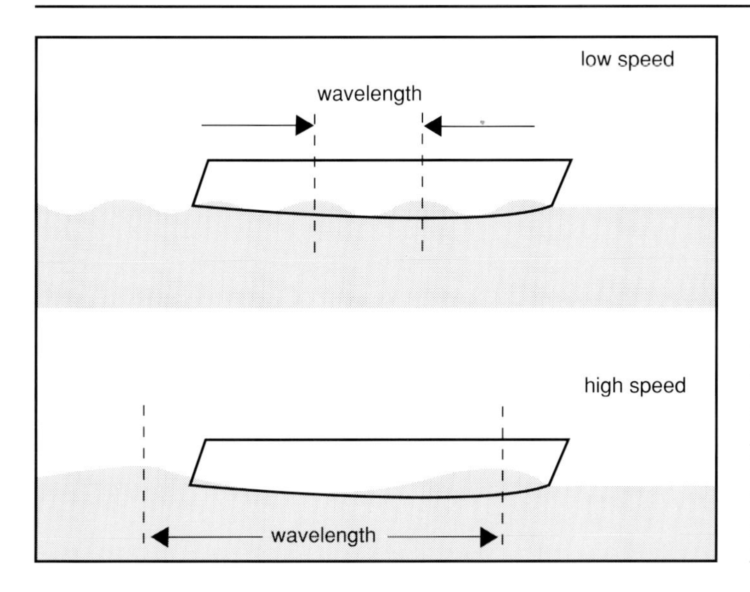

◀ *At low speeds the boat generates a series of small waves that create very little drag. As the speed increases the waves get higher and the wavelength longer, increasing the drag. When the wavelength is roughly equal to the waterline length of the hull the boat cannot go any faster unless it starts to plane.*

But waves have another, critical influence over a boat: they limit the top speed which any hull is capable of achieving in the displacement mode. As a boat increases speed, the waves it makes will vary in length. At its top displacement speed a boat will make just two waves along its length: the bow wave and the stern wave. Unless the boat goes over the 'hump' (rises over its bow wave) and starts planing it cannot go any faster.

The maximum length of the wave any boat can make depends on its prismatic coefficient (more later) and is approximately equal to the boat's waterline length. The wave itself has a finite top speed, which can be found by multiplying the square root of the waterline length (in feet) by 1.34. Therefore the maximum displacement speed of a boat with a 30-foot waterline is around 7.66 knots.

The eddy-making resistance is a resistance caused by the swirling water in the wake of the boat.

And then there is the drag developed by the keel and rudder as they develop lift. This is known as induced drag.

If a boat is travelling dead downwind its foils are not generating any lift, but they are still creating drag – frictional drag and section drag. As soon as the boat is turned away from a dead run its foils are required to create lift to prevent it being blown downwind. When a foil creates lift the water on one side of the foil is at a higher pressure than the other, so there is a tendency for the water on the high-pressure side to escape to the low-pressure side by passing over the tip. This not only reduces the efficiency of the foil but creates a swirling at the tip known as a tip vortex. These tip vortices create a resistance known as vortex drag.

The degree of drag induced by tip vortices depends on the area of the foil that they affect. Assuming the same total foil area, the longer the chord of the foil tip relative to the depth of the foil the greater the vortex losses – both in total and as a percentage of

▼ *The aspect ratio of a rectangular foil is calculated by dividing the span (s) by the chord (c). For an irregular shape the aspect ratio is the span squared, divided by the area.*

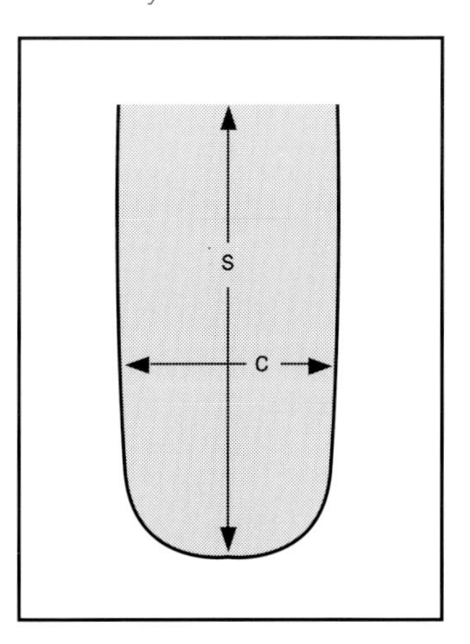

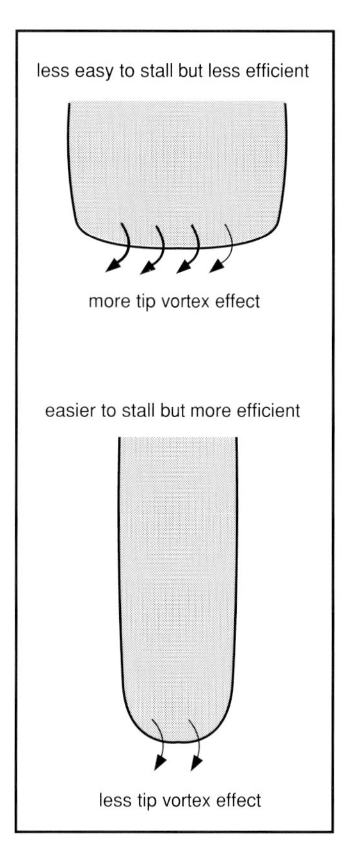

▲ *Foils of the same total area but different aspect ratios have different characteristics.*

the foil area. Up to certain limits, therefore, a shallow long-chord keel is less efficient than a deep short-chord one.

The relationship between the chord length and the depth is known as the aspect ratio. In general, the higher the ratio the better, to minimise the effect of tip vortices and improve the lift-to-drag ratio. The limiting factor, apart from the practical draft of the boat, is the stalling properties of the foil: a short-chord foil will stall more easily than a long-chord foil.

And also, as the yacht heels, there is the drag caused by heeling. As a yacht heels a component of the wind's force acts vertically downward which slightly increases the displacement and thereby increases drag (conversely if the rig can be canted over to windward, as with dinghies and windsurfers, a component of the wind force acts upwards, reducing displacement). When a yacht heels excessively not only do the rig and keel lose efficiency, but the immersed hull shape becomes very different – and usually less efficient – from the shape that the designer conceived.

STABILITY

So, we have an aerofoil – the rig – exerting a force on the boat, and a hydrofoil – the keel and rudder, and to some extent the hull – converting the exerted force into forward force. Meanwhile the boat is creating resistance that has to be overcome by the driving force. There is one other significant element, and that is stability.

Because the force of the wind acts at some distance above the water (it can be said to act at about a third of the mast height above the deck) and the force of the hull and foils acts below the

water, there is a tendency for the boat to roll over onto its side and go nowhere. So if the boat is to sail there must be a force preventing it from falling over, and that is known as the yacht's righting moment. In general, the greater the righting moment the greater the ability of the boat to convert wind power into forward motion.

Many people believe that the stability of a boat depends solely on its keel weight, but the truth is more complex. Stability depends on two elements: the weight of the boat and the relative positions of the boat's centre of gravity and its centre of buoyancy.

The stability of a boat is a moment – or leverage, if you like – composed of the downward force exerted by the weight of the boat, multiplied by the leverage created by the horizontal distance between that downward force and its equivalent upward force, which can be said to act through the centre of buoyancy. (The downward and upward forces are the same because the boat is floating; if one were greater than the other the boat would either sink or rise up into the sky!)

While the position of the centre of gravity (CoG) remains fixed as the boat heels (assuming the crew remain stationary and no-one is illegally piling heavy weights into the weather bunk!), the centre of buoyancy moves to leeward. This creates an increasingly large leverage between the two opposing forces and therefore the righting moment increases until it balances the heeling force of the rig. On keelboats, which cannot be prevented from heeling, this affects the way the hull is designed: the wider the boat, the further and more quickly the centre of buoyancy moves to leeward, and the greater the righting moment. On dinghies, where the crew can be moved to keep the boat is upright, it is not necessary to make the immersed sections of the hull wide to improve stability.

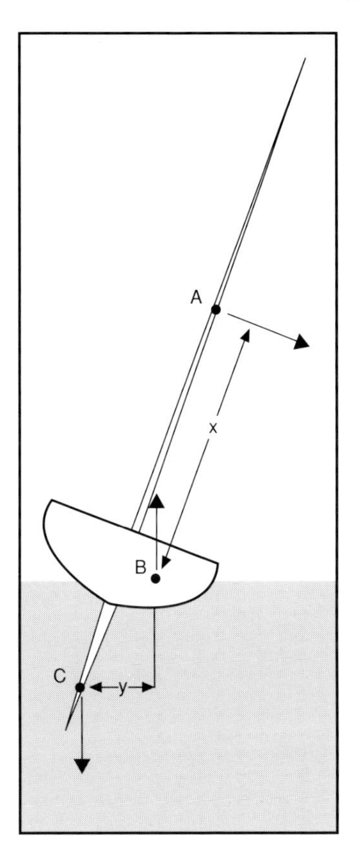

▲ *The overall force of the sailplan can be said to act at the centre of effort (A), a distance x from the centre of buoyancy (B). The weight of the boat can be said to act through its centre of gravity (C), a distance y from the centre of buoyancy. At a steady angle of heel, Ax = Cy.*

3 The hull

What functions does a hull perform?

It goes without saying really that a hull must float! But as well as floating, it must also offer the least resistance to forward motion possible in as wide a range of conditions as possible. It must provide a rigid platform to support the rig and the forces imparted by the rig and the deck gear, and in some cases it must contain the accommodation necessary to meet the class rules and racing criteria. In short, the requirements of a hull can vary widely depending on the type of boat, which may range from a flat-out day racer that is never afloat for more than a few hours to a world-girdling floating home carrying accommodation, stores and living facilities.

What affects the shape of a hull?

Many factors affect the hull shape, the most important being its application. Applications vary widely from dinghy racing on inland waterways to trans-ocean racing, and the more specific the application the easier the design brief. However, most designs are a compromise to get the best performance in a wide variety of conditions or to meet the criteria of a particular handicapping rule.

What are the important factors?

The hull of a boat can be described by a number of basic parameters – although it is possible for two boats with similar parameters to be vastly different.

Length The length is the fundamental proportion upon which all the others are based. For hydrodynamic purposes, the waterline length is the most significant length measurement because it determines the top speed a hull is capable of in its displacement mode – in other words the top speed attainable without planing. However, overall length is a commonly used dimension, particularly by marina operators!

Displacement Essentially, this is the weight of the boat. Strictly speaking displacement is a measurement of volume – the size of the hole the boat makes in the water – but this can easily be converted to weight by multiplying the size of the hole by the density of the water the boat floats in: salt or fresh.

Beam The beam – maximum and waterline – is generally proportionate to the boat's length and displacement, although narrower designs generally perform better upwind while wider designs generally perform better downwind. The difference between the maximum beam and the waterline beam depends upon similar considerations. For instance, if water ballast is to be carried or high crew stability without excessive waterline beam is needed, the hull might be flared in section to give a marked difference between the waterline beam and maximum beam. Conversely, if much of the boat's stability is derived from the keel the deck beam might not be signficantly greater than the waterline beam.

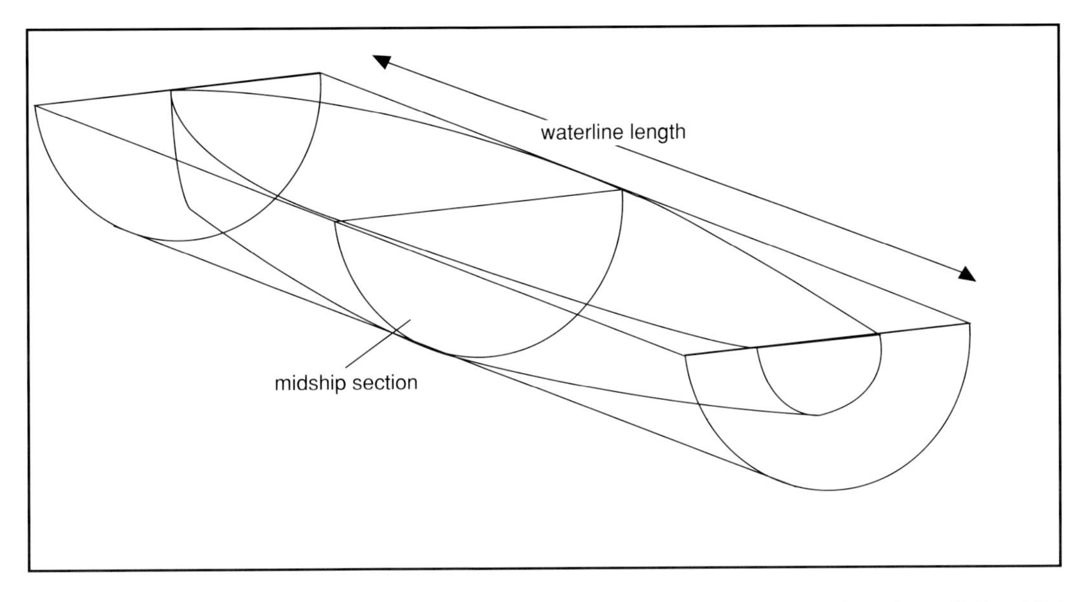

▲ *The prismatic coefficient (Cp) is expressed as the volume of the hull divided by the volume of a box of the same length and with a section matching the hull midship section.*

Draft This is measured either to the tip of the keel or to the bottom of the canoe body (the hull without appendages). For the purposes of hull shape and any parameters based on it, the canoe body draft is the most significant. For performance considerations and practical purposes the total draft is more significant.

Prismatic coefficient The prismatic coefficient (Cp) is a representation of how the displacement of a boat is distributed along its length. The formula for Cp is:

$$\frac{\text{displacement (m}^3\text{)}}{\text{Am} \times \text{lwl}}$$

In other words, the volume of the boat divided by the volume of a box with the same length as the boat and a constant section matching the boat's midship section (Am is the area of the mid-section). The prismatic coefficient has a fundamental bearing on the wave resistance of a hull and gives an indication of how the volume of the boat is distributed: a boat with a high Cp will have a relatively smaller midship area than one with a low Cp. In practical terms a yacht with a low Cp will be faster in light airs because there is less 'frontal area' presented to the water and a boat with a high Cp will be faster in a breeze. Additionally, a boat with a low Cp – and thus finer ends – will sink into its wave system more easily at high hull speeds, increasing resistance at the higher speeds.

Block coefficient The block coefficient (Cb) is the relationship between the volume of a boat and the volume of a box with the same length, beam and depth (see overleaf). The formula for Cb is:

$$\frac{\text{displacement (m}^3\text{)}}{\text{length} \times \text{beam} \times \text{depth}}$$

Using this formula, a brick would have a Cb of 1.

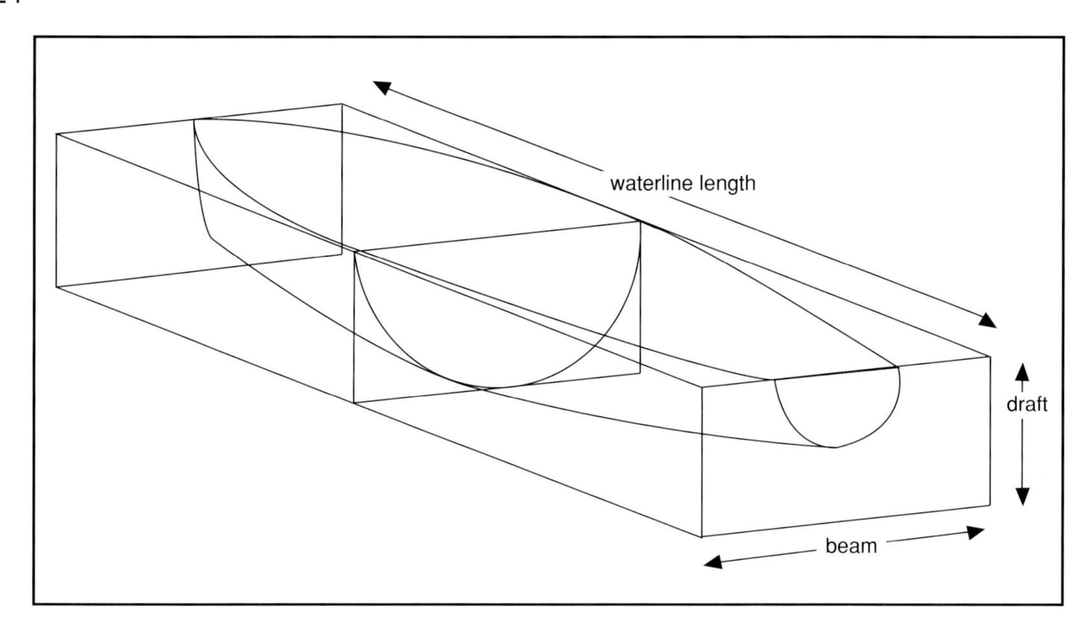

▲ *The block coefficient (Cb) is the relationship between the volume of the hull and the volume of a box with the same length, beam and depth.*

Tonnes per centimetre The tonnes per centimetre (TPCm) figure is a measure of the load-carrying ability of a boat, and is basically a definition of how much weight must be added to the boat to sink it by a centimetre. It is calculated by multiplying the area of the waterplane by one centimetre, and converting that volume to a weight by multiplying it by the density of the water in which the boat is sailing. A multihull with narrow hulls has a lower TPCm than a monohull, which is why multihull performance suffers more markedly from carrying excess weight.

Righting moment This is a measure of the stability of a boat, and is expressed in kilogram metres. The righting moment of any yacht is a measure of its tendency to return to the upright position when it is heeled over. In a sailing boat this is of fundamental importance since the figure dictates how much sail a boat can carry and therefore how powerful the 'engine' (sails) of the boat can be. A boat which heels easily has a lower speed potential than a similar boat which heels less or is 'stiffer'. There is an advantage in being tender, however, since in light airs the boat will pitch less in waves.

A boat's righting moment varies with its angle of heel. Unless it is fitted with a ballast tank or the crew are sitting off-centre, the righting moment when the boat is upright is zero. It increases as the boat heels until it reaches a maximum, usually (but not always) at around 90 degrees of heel. It then diminishes until it reaches zero again as the boat inverts, after which most boats have negative stability (when they have turned turtle they stay that way!)

For racing purposes the stability above 30 or 40 degrees is of little significance until seaworthiness becomes an issue. But if strong gusts or large waves tend to flip the boat to large heel angles, the wider and more efficient racing shapes become more prone to turning turtle.

RIGHTING MOMENT

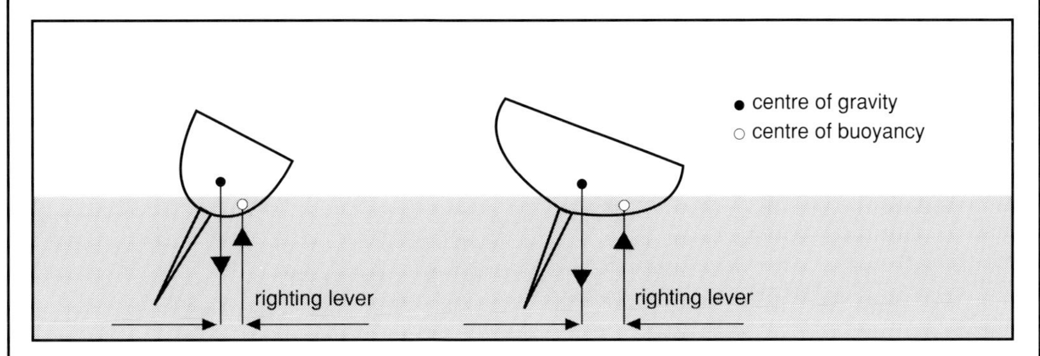

As a boat heels its centre of gravity remains static while its centre of buoyancy moves to leeward. The distance between the two is the righting lever, or G-Z. As the boats heels the G-Z changes, and the amount it changes depends on the shape of the hull: on a wide, flat hull the G-Z increases rapidly, and at small angles of heel such a hull has a greater righting moment than a narrow hull. At very large angles of heel, however, a wide flat hull has a greater tendency to turn turtle.

The righting moment of any hull can be plotted on a graph known as a G-Z curve, which shows not only the stability of the boat when it is the right way up, but also when it is capsized. Note that while a wide, flat hull has higher initial stability it soon diminishes; a narrow hull retains positive stability for a lot longer, and has relatively little negative stability when inverted.

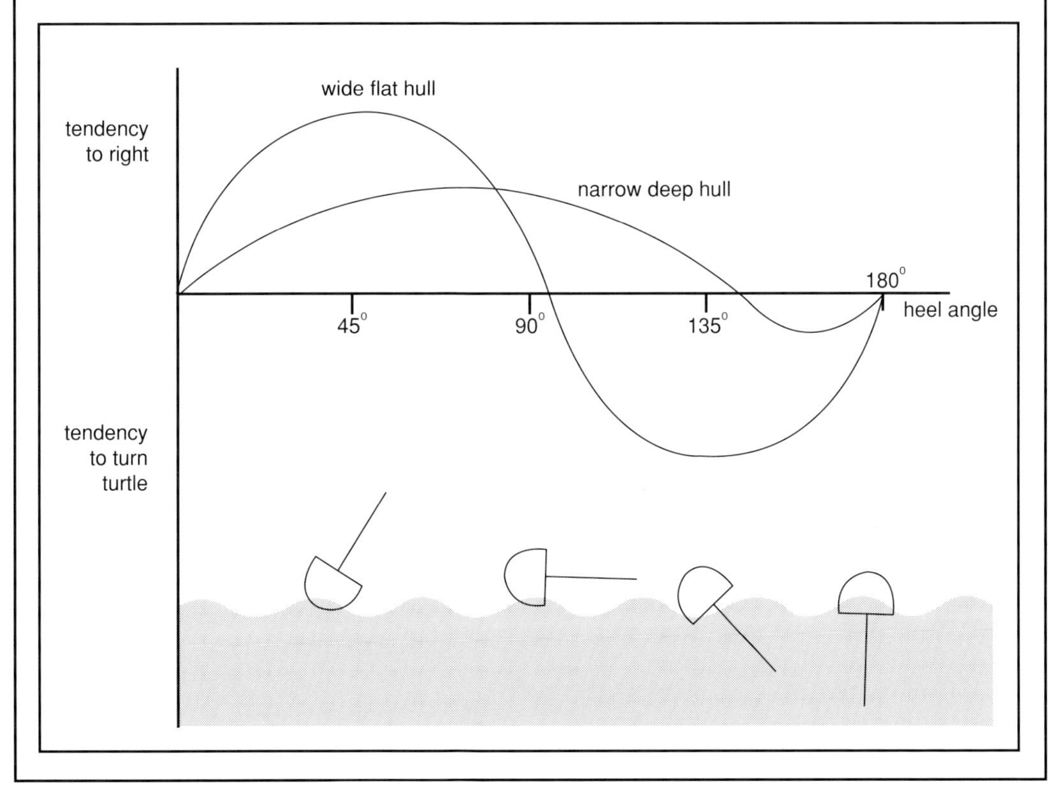

▲ *The straight stem seen on this boat maximises the waterline length compared to the overall length. The drawback of a straight stem is that it offers little scope for reserve buoyancy in the bow sections; this can lead to a tendency to nosedive.*

What is the relationship between overall and waterline length?

Overall length is a convenient approximation of the size of a boat but in terms of speed it is less significant than waterline length. In classes where overall length is restricted but waterline length is not, designers will try to create a boat with a waterline length that is as near as possible to the overall length. However, this does have a significant effect on hull shape, particularly at the bow: a vertical stem gives either a bow section that is not flared, or makes severe distortions necessary in order to maintain some flare. Bow flare is desirable because it helps reduce pitching by increasing the area of the waterplane as the bow is immersed.

What factors affect potential hull speed?

Hull speed varies with the conditions, but in the displacement mode it is limited by waterline length. At speeds of less than the maximum hull speed there are several factors which affect how fast a boat travels:

Beam/draft and stability The narrower the beam and the smaller the draft (in fact the smaller the midship section) the more easily a boat will travel through the water. However, beam gives a boat form stability and so a boat with a narrow beam will heel more easily and will not be able to carry the power of the rig as easily as a wider boat. The optimum beam is therefore a trade-off between stability and ease of movement through the water.

Wetted surface area This affects the friction drag on the boat since the higher the area, the greater the surface the water has to pass over. In general, wider boats have higher wetted surface areas than narrow boats, but wetted surface area is traded off against other factors such as stability and flat aft sections which help promote surfing and planing.

Displacement The weight of a boat does not actually affect its potential top speed but it does affect the acceleration and how much force is necessary to attain a particular speed. The heavier a boat is, the slower it will accelerate – but conversely the longer it will carry its way when the wind drops. Sometimes a heavier hull pays when sailing upwind in wind speeds at which the hull is travelling at hull speed. Firstly, as the boat is lower in the water its waterline length is increased and so, therefore, is the maximum hull speed (max hull speed in knots = $1.34 \times \sqrt{\text{waterline length in feet}}$). Secondly, as the boat is heavier and stability depends on the righting moment (which is a function of the weight), stability will increase. However, heavy boats are slower to plane and surf and so downwind it always pays if the hull is as light as possible.

Wavemaking The wavemaking of a hull depends upon its weight and shape. A heavy boat will make more waves than a light boat and some shapes will make more waves at certain speeds than others. The prismatic coefficient is an indicator of this. The designer's job is to tune the wavemaking characteristics of a design to its envisaged sailing conditions to ensure that the proportions of the hull minimise wavemaking for the desired weight, beam, draft and length. For a given set of these dimensions there are many shapes of hull with widely differing wavemaking potentials.

Trim and heel also have a profound effect on wavemaking, so when you are experimenting with moving weight fore and aft what you are really doing is attempting to minimise the wavemaking of the hull by preventing the stern from dragging or by presenting the sharpest section of the hull to the oncoming water. As far as heel is concerned, most dinghies have flat slab-sided topsides in the bow area so that when they heel the entry angle of the bow is widened and a flat face is presented to the water – which is slow. In contrast, since keelboats cannot be sailed upright they are often designed so that the best bow sections are presented to the water when the boat is heeled slightly.

What are the elements of a boat's hull?

A modern hull consists of an outer skin, which maintains the watertight integrity, plus a framing system of some kind. The framing system usually fulfils two roles: it supports the hull skin and enables the skin to maintain its designed shape; it also supports the loads imparted by the rig, keel, rudder and crew. In essence, the ideal internal structure prevents fore-and-aft bending and twisting, and allows enough athwartships rig tension to be imparted to hold the rig straight in the boat without distorting the hull. However, most framing systems are a compromise, because internal fittings and accommodation are usually required in keelboats and weight is also a consideration.

What is the significance of hull weight?

Designers and builders go to great lengths to reduce hull weight. It is necessary to keep hull weight to a minimum for two reasons. Firstly, for a given all-up weight, any weight that is saved in the hull can be used for ballasting the boat which lowers the centre of gravity and increases the righting moment. Secondly, a heavier hull is more prone to pitching.

What affects pitching?

The amount a boat pitches – and thus dissipates energy from the rig and reduces the efficiency of the rig – depends upon what is known as the radius of gyration of the boat. This is not a simple radial measurement , but a measure of the distribution of the weight about what is known as the 'quiescent point', or the point about which the yacht pitches. On a rotating wheel the quiescent point is at the centre. A pitching yacht has a similar point which is roughly the centre point of the waterplane.

The radius of gyration depends upon the weight of every item on the boat and the *square* of its distance from the quiescent point. So although the centre of gravity of a heavy mast may be the same distance from the quiescent point as that of a light mast, the extra weight will increase the radius of gyration. Similarly, the anchor of a yacht placed in the bow will have the same weight but a greater radius than an anchor stowed near the middle of the boat. It therefore pays to stow all movable weight amidships, keep the hull – and indeed the rig and the deck fittings – as light as possible, and if a class rule stipulates a minimum weight, correctors should be added as near to the middle of the boat as possible.

Does the keel add to the pitching problem?

Indeed yes. Modern boats with deep keels carrying heavy bulbs on the end suffer far more from pitching problems than boats with shorter keels and ballast in the keel or internal ballast. Pitching can be suppressed by wing keels but they sap vital driving-force energy in so doing. No system of suppressing pitching is as effective as reducing it in the first place so, where the rules allow, concentrating weight in the form of ballast, tools, sails and anchors will always be faster.

How important is hull finish?

As discussed earlier, there are two elements to the surface of a hull: its fairness and finish. Fairness is dealt with in the next section.

Hull finish is more important in light winds than in heavy winds because the frictional resistance – to which the surface finish is relevant – is a greater proportion of the overall resistance at slow speeds. Given two equal boats, the boat with the best surface finish will always be faster.

For most purposes, the smoother the finish of a hull, the faster it will be. Merchant ships have experimented with various polymer fluids ejected from the hull at the bow which alter the properties of the water as it flows over the surface of the ship. Claims of a 30 per cent reduction in skin friction resistance have been made, and there is no doubt that such systems do work. However, they are outlawed in yacht racing and in any case are highly harmful on the environment. For racing purposes, 600 grade sandpaper is the most efficient answer where an antifouling is used, and a high polish where there is no antifouling.

What effect do hollows, bumps and protrusions have on boatspeed?

The fastest hull is a smooth hull with a fair shape around which the water can flow. The precise shape of the optimum hull depends upon the speed at which it travels through the water but in general any protrusion, bump or hollow will cause drag by virtue of its own shape and disturb the flow of the water to create speed-sapping turbulence – particularly at the stern where the water is separating from the hull and joining the boat's wake.

Hollows, bumps and protrusions are generally made necessary by some kind of handicapping rule; the designer will have calculated the increase in resistance and traded that off against the gain in handicap. Unintended hull deformations can be highly detrimental to speed and should be eliminated. Skin fittings, speed sensors, rudder flaps and joins between hull and keel, or rudder and skeg, should all be faired as closely as possible.

Hollows and ripples often occur in poorly-built boats where the hull laminate has distorted and created an unfair surface. Such problems can be overcome with filler and fairing compounds; the gain in performance usually offsets any added weight of filler.

How important are trim and heel?

Trim and heel are of vital importance. They alter the underwater shape of the hull and affect the shape of the waterlines, the wetted surface area and the amount of helm needed to steer the boat. They vary from boat to boat, and they vary with wind and sea conditions. Boats with narrow waterlines and flared sections generally need to be sailed more upright than boats with less flare. Boats with flat bow sections (i.e. a flat bottom near the bow) are generally faster trimmed with the bow 'pinned down' – trimmed forward to reduce slamming into waves. However, the ideal heel and trim for a keelboat can only be found through practice and tuning.

▲ *Sailing a dinghy upwind with weather heel enabes the weight of the rig to add to the stability rather than detract from it, and also lifts the boat out of the water slightly. The technique certainly improves boatspeed.*

▶▶ *It is rarely possible to sail a keelboat dead upright, let alone with weather heel. However, in anything other than the lightest of airs the boat should be sailed as upright as possible by piling the crew on the weather rail. Here the boat is heeled about 10 degrees too far, and needs more weight up or less power from the rig.*

Most dinghies are designed to offer least resistance when sailed upright. In this condition a good design will carry a slight amount of weather helm, enabling the rudder to contribute to the lift generated by the keel. However, in some circumstances it pays to sail a dinghy upwind with some weather heel because in this condition the weight of the rig is actually contributing to righting moment rather than heeling moment, and there is also a very slight upward force lifting the boat out of the water rather than pressing it down.

In a boat like a J/24 it pays to sail as upright as possible – which is why a heavier crew is usually faster. This is because a J/24 has a small keel and as it heels the keel loses efficiency. In a narrow, heavy boat like a 6-Metre or 12-Metre the crew weight makes very little difference to the degree of heel and the boat is not slowed in the same way by heeling.

It is usually impossible to sail a keelboat upright in anything other than the lightest conditions since the keel will not provide any righting force until some heel is generated. However, it is usually fastest to sail with a heel angle of less than 20 degrees, since at greater angles the underwater shape of the hull is distorted and offers extra resistance to motion; the keel also loses efficiency. The maximum desirable heel varies from design to design but, in general, a narrow boat will be slowed less by heel than a wide one.

When reaching in both dinghies and keelboats it is sometimes desirable to sail with a small amount of leeward heel since this will usually eliminate any weather helm and therefore reduce the resistance of the rudder. In a 505 or an Ultra, where the spinnaker pole is very long, the boat has considerable lee helm when sailed upright on the reach. If the boat is heeled slightly it will track dead straight without using the rudder. Similarly, dead downwind it is often fastest to heel a boat to windward; this neutralises the helm by bringing the centre of effort of the rig over the centre of resistance of the hull.

Trim is more dependent upon the amount of wind and is more easily controlled by the crew. In general, the wider sections towards the rear of a boat create greater wetted surface area than the narrower sections at the bow of a boat. Trimming a boat down by the bow lifts the stern out of the water and therefore reduces the wetted surface area; this reduces the resistance in light airs when the hull is not moving at speeds approaching its hull speed. However, as boatspeed increases, sailing length becomes more important and it is usual to trim the boat further down by the stern to gain the maximum waterline length.

The amount of trim required at any particular time can only be found by trial and error, but in general a boat should be trimmed down by the stern (while sailing) until the stern wave is running cleanly off the transom. If the water appears to be spurting up from under the stern the boat will be dragging a large stern wave.

In light airs a combination of trim forward and heel to leeward will reduce the wetted surface area and generate some feel in the helm which will make the boat easier to steer efficiently.

GBR
GBR
180R
180R

▶ *Moving the crew weight forward when sailing upwind keeps the bow in the water and reduces slamming.*

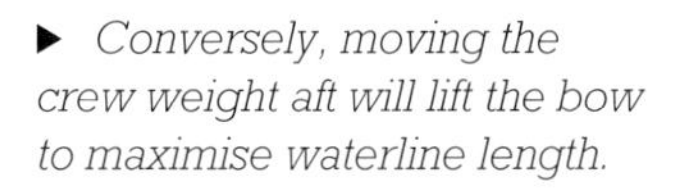

▶ *Conversely, moving the crew weight aft will lift the bow to maximise waterline length.*

Why trim forward upwind and aft downwind?

Sailing upwind generally involves sailing into waves which tends to result in the boat slamming. This absorbs energy which might otherwise be converted into forward motion, as well as inducing shock loads into the keel and rig which affects the flow over them and reduces their efficiency. Moving weight forward keeps the bow in the water and helps prevent slamming.

Downwind a boat is generally sailing with the wave system, and must be trimmed aft to prevent the bow burying. There is also a large bow-down trimming moment created by the rig pushing forward, especially when a spinnaker is set, which must be counteracted by moving the crew aft.

What makes for low and high wetted surface area?

Differently shaped solids of identical volumes can have widely differing surface areas and a designer is always trying to devise desirable shapes with the required volume while minimising the surface area. A semi-circular hull section gives the lowest possible surface area, while at the other end of the scale a flat, shallow section gives the maximum area. For other hydrodynamic reasons, boat hulls generally approximate to semicircular sections forward and flat, wide sections aft.

▲ *Flat aft sections help promote planing and surfing.*

Why?
Because the teardrop-shaped waterplane offers the least resistance with a narrow angle of entry, a gentle flare to the widest point and a gradual tapering off.

In displacement conditions the optimum hull would taper to a thin tail, but downwind – when planing or surfing is desirable – wider, flatter sections are needed. Narrow stern sections also make a boat very difficult to steer downwind since there is very little 'form' stability provided by the shape of the hull to prevent rolling. This problem is particularly acute with modern short-chord keels which themselves offer little directional stability.

What is the conflict between an upwind shape and a downwind shape?
In general, upwind sailing takes place in the displacement condition, and so the fastest shape is one which offers the least resistance to the water passing over it. A narrow angle of entry, flaring out to a moderate maximum beam and smoothly tapering back to narrow exit offers the least resistance. However, upwind boats usually sail at some angle of heel and so the actual upright waterline shape is not as significant as the waterline shape of the heeled hull which should be similarly fair.

The downwind condition is complicated by the fact that a hull operates both in the displacement mode and in planing/surfing mode. In the displacement mode a fair shape is important. However, form stability is more useful downwind when, under spinnaker, the heeling force is greatly increased and wide aft sections prevent excess heel. A stern with wide, flat sections will also plane or surf more readily than a narrow one.

What is surfing and planing?

There is no actual difference between surfing and planing as far as the action of the hull is concerned. In both cases the boat overtakes its wave system and rides on the water rather than in it, which enables it to travel faster. The difference lies in what causes the boat to rise up: planing is induced by the force of the wind while surfing is caused by the boat travelling down the face of a wave. It is possible for a boat – particularly a dinghy – to surf without the aid of the wind but in general, on keelboats, surfing is a combination of wind and wave forces.

Most boats will surf in the right conditions but only light boats with flat sections will plane.

What is leeway?

Leeway is the difference between the direction the boat is pointing and the direction it is travelling. Boats make leeway because, in order to work, their keels must be set at an angle of attack to the flow of the water – and since the keel is attached in line with the centreline, the whole boat must be set at an angle to the flow. It is an angle that cannot be eliminated, except dead downwind when the keel is not acting as a lifting surface.

Reducing leeway is a major preoccupation of the designer because the hull itself offers least resistance when it is aligned with the water flow. Various methods have been tried, including swinging keels that can be set at an angle to the hull so that the keel maintains its angle of attack while the hull moves straight through the water, but such devices have met with limited success.

There are various ways that the leeway angle can be deduced: one of the simplest involves dropping a floating (biodegradable of course!) object over the transom and comparing its angle, as the boat sails away, with the reciprocal of the angle being steered. A similar technique involves comparing the angle of the boat's wake with the reciprocal of the course steered.

The more upright a boat is sailed the smaller the leeway angle, but there is very little that can be done to alter the natural leeway angle of a boat apart from altering the size and shape of the foils, which in some classes is prohibited. That said, if a boat is making more than about eight degrees of leeway upwind there must be a problem with the design of the boat or the condition of the foils. Experimenting with sailing lower and faster, and higher and slower, should determine an optimum speed at which the keel is working efficiently in terms of its angle of attack (the leeway angle) and the speed of the water flowing over it.

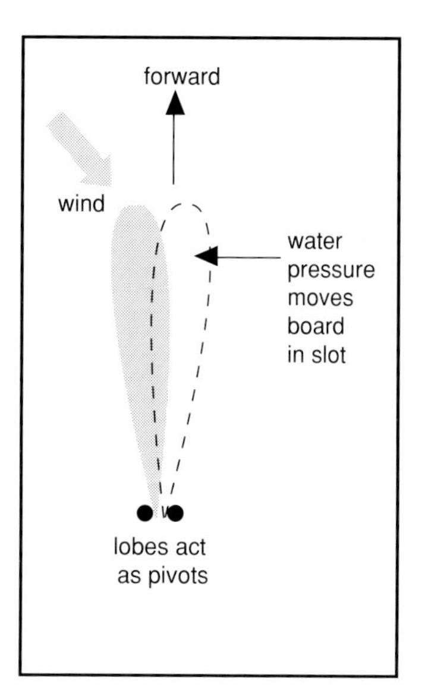

▲ *The principle of a pivoting keel (or gybing centreboard on a dinghy) is that as the keel takes up its natural angle of attack (the leeway angle) the hull is effectively pivoted over the top of the keel and therefore aligned with the actual direction of travel, reducing drag. These devices often work better in theory than in practice.*

4 Hull appendages

Hull appendages take many forms and have been the subject of intensive research and development as yacht design has evolved. Designers have concentrated on making keels smaller, more efficient and more versatile; and rudders more powerful, less likely to stall and with less drag.

There have also been developments such as winged keels and canards (daggerboards in front of the keel) which have proved highly effective at international level in specific circumstances.

What is the difference between a keel and a centreboard?

A centreboard provides no stability, just hydrodynamic forces; a keel provides both hydrodynamic forces and varying levels of stability. For the purposes of this chapter we will consider the keel.

How does a keel work?

A keel has two functions: it provides stability to a keelboat, and it provides a sideforce in opposition to that generated by the sailplan. We covered stability in Chapter 2, but here we can enlarge on the way the keel provides lift or sideforce.

If you put your hand out of the window of a car while it is moving, your arm experiences two of a number of forces. There is always a backwards force and, depending upon the angle of your hand, there can be an upwards force, a downwards force or no other force. In this situation your hand represents the keel, with the upward and downward forces representing the side force on either tack and the backward force representing the drag. In the horizontal position your hand represents a keel dead downwind when the sideforce is zero and the only force is the drag force.

Theoretically, the forces acting on a keel can be divided initially into two components – lift and drag. Lift is described as the component from the keel that opposes the force from the sails and creates the driving force, while drag is a component that acts in opposition to the direction of travel of the yacht. For efficient sailing the lift force must be maximised and the drag minimised. The efficiency of a keel is determined by its lift/drag ratio.

What creates drag?

The total drag or resistance of a keel has several constituent parts. Firstly there is the friction drag, which is a function of the surface area; the bigger the keel the higher its frictional resistance. Related to the friction drag (in that it is a function of size) is the section drag, which varies according to how the shape is pushed through the water. Then there is the induced drag, which is the cost of the keel providing lift. And finally the vortex drag is caused by the escape of water from the high-pressure to the low-pressure side of the foil: from the leeward side to the windward side.

What are the significant factors?

There are several important factors that determine the design and efficiency of a keel.

Lateral area When combined with the lateral area of the hull and other appendages such as the rudder, this must be sufficient to provide the necessary sideforce for any envisaged sailing condition.

Planform This is the shape of the keel in profile (viewed from the side). The planform has a profound effect on the way the water flows over the keel, particularly at the tip where the greatest loss of efficiency occurs as the water escapes from the high-pressure side to the low-pressure side. In an ideal situation the water would flow evenly (horizontally) over the keel and there would be no loss of pressure from one side to the other. In practice this cannot be arranged, but minimising the losses known as tip losses is one of the main aims of advanced keel design.

Aspect ratio The aspect ratio is a function of the planform and the area. For a rectangle it is the ratio of the depth to the chord length but for more complex shapes it is the depth (span) squared divided by the area. Up to certain limits the higher the aspect ratio, the higher the lift/drag ratio – and therefore the higher the efficiency of the keel. However, a high aspect ratio has certain drawbacks

such as earlier stalling and a greater loss of efficiency due to pitching and uneven flow.

Section shape This can be compared to the wing shape of an aircraft, and has a fundamental effect on the lift of a keel. However, it is limited by the fact that the section must be symmetrical in order to work on both tacks. This affects the nature of the section shape and effectively limits the range of sections that can be used. In the case of a ballast keel there is also a certain volume of ballast material – usually lead or iron – to be carried and this can affect the nature of the section chosen.

End plates or bulbs These are designed to reduce the tip losses described above by acting as a kind of 'fence' to the water which is trying to escape around the end of the keel. They usually take one of two forms: either a bulb, which is a hydrodynamic torpedo, usually containing some ballast to allow the fin of the keel to be smaller for a given righting moment; or some type of wing which may contain ballast but which is designed to offer an efficient hydrodynamic resistance to leeway as well.

What is lateral stability?

Lateral stability is the tendency for a boat to track straight rather than weave or even broach out of control. The longer the keel extends fore and aft (chord length) the more laterally stable the boat; it will track straighter but will also be more difficult to turn. Modern short-chord keels are harder to keep travelling straight; on the other hand they are easier to turn and are more efficient.

How does the section shape of a keel affect its efficiency?

Unlike the wing of an aeroplane which is required to provide lift in only one direction – upwards – a boat's fin must provide lift in two directions if the boat is to sail efficiently on both tacks. In order to provide lift, a symmetrical foil must be set at an angle of attack to the flow. This is the leeway angle of the yacht and occurs naturally.

The section shape of the keel is designed to give lift on both tacks with minimum drag. However, because the keel works at both low and high boatspeeds, and in both waves and flat water, its section is not a high-efficiency profile for specific conditions but the best compromise for a range of conditions.

Are all keels compromises?

Most are, because most are constrained by draft, a range of conditions, or weight. For races like the Whitbread race, where weather data has shown that the prevailing sailing condition is reaching in moderate winds, the boats are designed with specialised reaching keels, with very low surface areas and fat, high-lift sections. They perform well when reaching but are very slow out of tacks and off a starting line. Similarly keels and centreboards for speed trials are designed to sail in one direction only at high speeds. In general, however, keels are designed to perform as well as possible in a wide range of wind and sea conditions.

And those wings and bulbs, do they work?

After the America's Cup of 1983 when *Australia II* won the cup with a cleverly-designed winged keel, every boat at every boat show had a winged keel. Unfortunately most of these were little more than fashion accessories. There is no doubt that wings and bulbs *can* work; they can reduce the tip losses, reduce the surface area and lower the ballast position. However, a successful wing or bulb is usually the result of hours of painstaking testing which is essential to determine the right proportions for the boat and the conditions, and most of the devices adorning the fashionable yachts of the mid-1980s did more harm than good.

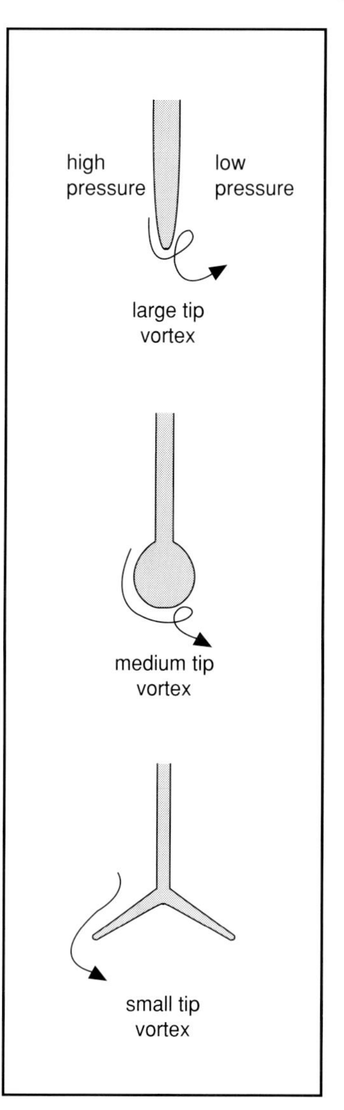

▲ *The more tortuous the route from the high pressure side to the low pressure side, the less inclined the water will be to make the journey. But tip appendages create drag, so their design is a trade-off of higher lift against higher drag.*

How is a keel designed?

When designing a keel the designer usually faces a number of constraints, and has to find ways of reconciling the various factors to create an efficient, yet practical and permissable foil shape. Draft is likely to be the first consideration. In the absence of a limit on keel draft a designer would almost certainly opt for a deep, high aspect-ratio profile which is the most efficient type of foil. Stability is likely to be the other constraint. The designer must decide what righting moment is required in the design and therefore how much ballast must be carried in the keel. Stability is often restricted by handicapping rules, so maximum stability is not always a prerequisite. If there are no handicapping rules the designer has to trade off keel weight, and hence stability, against all-up weight and reach a satisfactory compromise.
End plates and trim tabs must be considered once the draft and the weight have been decided upon. If an end plate is to be used some of the ballast may be carried in the plate which will affect the nature of the rest of the keel design. If a trim tab is to be used this will affect the section choice, the required area and the shape of the profile.
Section, lateral area and planform must be considered together as they are all interrelated. Both the section shape and the planform have an effect on the efficiency of the fin, and a higher-efficiency shape will require less area for the same lift than a lower-efficiency shape. Because the foil is symmetrical the designer can choose from only a small range of sections and he must bear in mind the ballast requirements of the keel. Once the section is chosen the planform and areas are matched.

What is the difference between section shapes?

Assuming a keel is to be a symmetrical section with no trim tabs or other section modifiers, there is a range of sections that may be used. When aligned straight with the flow a flat plate with a pointed leading and trailing edge would give the lowest drag, but this is no longer the case if the keel is required to provide lift .

Starting at the leading edge, the narrower the foil, the narrower the band of angles of attack it can accommodate. A fat leading edge is more forgiving in waves or out of tacks when the boat is slipping sideways and the leeway angle is high.

Moving aft, the fatter the section the more power it can generate (up to a limit) because the greater the difference in distance along the leeward and windward surfaces. This gives the potential for greater lift but also greater drag when sailing dead downwind and the section is simply being dragged through the water. Therefore a fat section has a higher potential for lift in some conditions but also a higher potential for drag under others.

The position of the maximum fullness is related to the attachment of the flow around the windward or low-pressure side of the foil as discussed in Chapter 2. It is also affected by the nature of the leading edge and the amount of camber (fatness) in the keel. However, the maximum fullness should never be further aft than 50 per cent and is usually no further forward than 30 or 35 per cent.

What developments have taken place in section and planform?

For general purpose use the keel is always a compromise and the range of profiles and section shapes is limited. However, for specific use a keel can be customised to a great degree. In the Whitbread, as mentioned earlier, a high proportion of the race is spent reaching and running and only a tiny proportion sailing upwind. Thus Whitbread keels have shrunk dramatically in size to reduce wetted surface area. Ballast is carried in a large low-drag bulb and the fin area is very small. The boats are therefore less efficient upwind but faster reaching and running.

Similarly, in the America's Cup where technological development is paramount, keel and rudder systems using multiple fins and several rotating parts combining keel and rudder are often tried to increase efficiency in specific conditions.

How is the position of the keel on the boat determined?

The keel, the rudder and to some extent the hull all have an effect on the sideforce that is generated to overcome the sail force. The point at which the overall sideforce of the keel, rudder and hull can be said to act is known as the Centre of Lateral Resistance (CLR). Similarly, the point at which the overall sail force can be said to act is known as the Centre of Effort (CoE). In order to maintain a certain amount of weather helm – which is generally held to be necessary for optimum upwind performance – the CoE must be located slightly behind the CLR for upwind sailing so that the boat has a tendency to 'round up' if the helm is released. This is known as the 'lead', and the rig and the keel must be positioned to give the correct lead.

How do I determine the lead?

The lead is a simple mathematical computation: determine the lateral areas of the keel, rudder and hull canoe body (the lateral area is what you see when you look at the boat side-on). Using an arbitrary point – the stem or the stern – multiply the areas of the indi-

vidual items by the horizontal distance of the centroids of each from the arbitrary point, add them together and then divide the result by the total area of the items added together. This will give a distance from the point selected.

Carry out the same operation on the sail plan, multiplying the areas of each sail by the horizontal distance of their centroids from the arbitrary point and dividing by the total sail area. The difference between the two distances is the lead which, knowing the waterline length, can be worked out as a percentage. Once the figure is known it is possible to decide on either mast or keel position or, if they are fixed, the amount of rake required in the rig to obtain the optimum lead. Roughly 12 per cent of the waterline length is generally thought the correct amount.

How much should a keel or centreboard be retracted for downwind sailing where possible?

This is largely a matter of feel. Theoretically, dead running requires no board area at all and the more it can be retracted the faster the boat should be. In practice, however, the board provides some damping of any rolling tendency and it might be very difficult to sail some boats downwind with the board fully retracted. When tacking downwind and sailing VMG angles some board will be necessary to stop the boat slipping sideways and losing power in the rig.

When reaching, the amount of board depends upon the weight of wind, the weight of the crew and at what angle it is required to sail. On a broad reach the board can be retracted until the boat feels balanced on its course and it is not tending to heel over too much. On a tight reach it might be necessary to use more board in order to lay a mark, accepting a greater angle of heel.

In general, apart from dead running, the amount of board needed can be deduced by the feel in the helm. If the board is retracted so far that lee helm is being generated then it is retracted too far. Lower the board until the helm feels neutral.

How should the trailing edge be finished?

In an ideal world the trailing edge of a keel should be infinitely thin, but this is obviously not possible. What's more, it is not even necessarily best to get the thinnest edge possible because the turbulence created behind a thin edge that is not infinitely thin will cause the edge to vibrate and lose lift. The compromise is a neatly squared-off trailing edge that is as thin as is practical. The main criterion is that the foil should not vibrate. If vibration continues after the foil has been faired and squared off at the trailing edge then the problem may be minimised by chamfering the edge at an angle rather than squaring it off.

How important is the surface finish of the keel?

The surface finish of the keel – and rudder – is of paramount importance: much more so than the hull surface. Since the keel and rudder are lifting surfaces, the smooth flow of water around them is critical to their performance and not just to the amount of frictional resistance they create.

How can I tune my keel?

In a one-design class where all dimensions are rigorously stipulated, there is very little that can be done about keel shape or position. However, it is essential that the sections of the keel match up to the desired aerofoil sections specified, so checking that the keel is not twisted, and is mounted straight – both vertically and longitudinally – is vital. You should also check the keel against section templates; make wooden templates from the plans, for several vertical positions, and offer them up to the keel. Note any discrepancies and fill or sand the foil as appropriate. Ensure that the surface finish, and the keel to hull joint, are as clean and smooth as possible.

In classes where some latitude is allowed, there are some modifications that can be made. Firstly, confirm that the position of the keel is correct, giving a comfortable amount of weather helm (five degrees) in medium conditions upwind. If not, the keel can be moved forward or aft to get the correct balance. Also, you can examine the planform and section of the keel together with the lateral area. If you are suffering poor upwind performance, the lateral area of the keel may be too small or the planform an inefficient shape. If the boat is slow when reaching and running the keel area may be too large or the thickness too great.

RUDDERS

A rudder, like a keel, is an aerofoil that provides lift and creates drag. As with a keel the lift/drag ratio is important but because a rudder pivots and is used to turn the boat there are other considerations, notably its angle of stall and its ability to accommodate a wide range of angles of incidence.

How does a rudder work?

In straight-line sailing all the forces from the rig are balanced by forces from the hull, keel and rudder. They are balanced about a point near the middle of the boat. Upwind the rudder is usually a contributor to the overall sideforce that keeps the boat balanced and sailing straight. This is weather helm.

Note that weather helm is not simply the angle of the rudder to the centreline but the angle of the rudder to the water flow. The amount of weather helm is the rudder angle added to the leeway angle of the boat, and even with the helm amidships a small amount of weather helm is present. In some boats, particularly dinghies, it is fastest to sail with a minimum amount of helm. This is because the keel or centreboard is large enough to provide the required sideforce and creating additional sideforce using the rudder only increases the drag.

If the rudder is turned to align with the water flow its contribution to the sideforce ceases and the boat becomes unbalanced and turns. If the rudder is turned further in the same direction it begins to create a force of its own which – within limits – increases as the angle increases. This exerts a turning moment on the boat.

Why should a rudder have weather helm upwind?

Ther are two reasons: firstly, it is very difficult – though not impossible – to steer a boat accurately with neutral helm. Physically, it is easier to steer with a slight force acting against the arm of the helmsman. Secondly, it is possible for the rudder to be used to create some lift in the same manner as the keel, by setting it at an angle of attack to the flow. Otherwise the rudder, unless turning the boat, simply creates a drag force and no other beneficial forces.

What factors affect turning?

Several factors affect turning: the speed of the boat, the amount the rudder is turned, its area, its efficiency as a hydrofoil, its distance from the CLR, the underwater profile of the boat and the trim of the sails.

▼ *Under normal sailing conditions of weather helm the rudder contributes to the side force (top) and keeps the boat tracking straight. The boat will naturally tend to round up if the helm is released. If opposite helm is applied the rudder will create a turning moment of its own (bottom).*

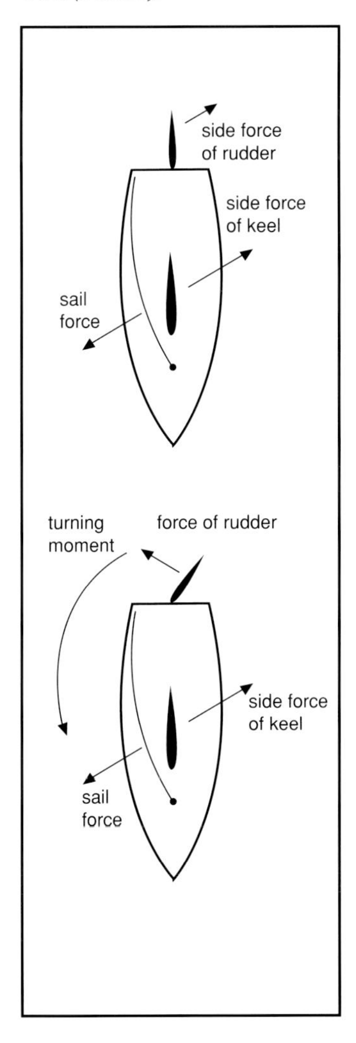

The lift force is related to the square of the speed of the water flowing past the rudder, so if the boatspeed is doubled the rudder force is quadrupled. This is why fast boats need much smaller rudders than slow boats.
The amount the rudder is turned is roughly proportional to the turning force it exerts up until the point it stalls. Depending upon its section and profile shape, the rudder stalls when the angle of incidence with the passing water is so great that the water can no longer 'cling on' to the low-pressure surface.
The area of the rudder is proportional to the force it is able to generate but also has an effect on the drag it creates under steady sailing conditions.
The aerofoil efficiency is similar to that of the keel except that the rudder often has to maintain large angles of attack without stalling. The same hydrodynamic principles apply in terms of section shape, aspect ratio and planform but the section shape is slightly thicker. Although the thicker section increases the drag it also increases the acceptable angle of incidence without stalling.
The distance from the CLR has a profound effect on the turning efficiency of the rudder because turning is caused by a moment which is the force of the rudder multiplied by its distance from the CLR. The further the rudder is from the CLR, the smaller its area has to be to produce the same turning moment. Therefore rudders are often hung off the transom rather than under the counter.
The underwater profile affects the speed of turning. A boat with a shallow hull and a deep keel and rudder will turn more quickly and be more responsive than a boat with a deep hull and a long keel because immersed lateral area resists turning. Conversely, a yacht which is slow to turn will usually maintain a steady course more easily than a more manoeuvrable boat.
The trim of the sails can have a dramatic effect. If the headsail is suddenly released, the CoE of the rig moves aft and a turning force into the wind is developed. Similarly, if the mainsail is released the CoE moves forward and a turning force away from the wind is developed.

Cooperation between the sail trimmers and helmsman will give fast or slow turning as required. When sailing a straight line, the balance of sail trim that requires the least movement from the rudder will be quickest.

What is the best rudder profile?

Because the rudder is required to be forgiving to a wide range of angles of incidence without stalling it must be of moderate aspect ratio. A deep rudder with a narrow chord (a high aspect ratio) generates high lift forces at low angles of attack, but also has a low stall angle – and a stalled rudder is effectively useless.

What is rudder balance?

Because the rudder is a moving foil, the vertical axis about which it turns affects how it behaves and how much weight is felt by the helmsman through the steering system.

◀◀ Andrew Preece powering the Mumm 36 to windward.

If the rudder was pivoted at its leading edge its turning axis would be some way from the centre of effort of the foil and the helm would feel heavy. If the rudder was pivoted at the CoE there would be no weight in the helm at all, even when the rudder was providing a huge force to turn the boat. If the rudder was pivoted at the trailing edge the rudder would be unstable and would be trying to turn continually. In practice the optimum balance is with the pivoting point somewhere between the leading edge and the CoE so that sufficient 'feel' is induced for efficient steering without overstraining the helmsman.

What are the effects of the length of the tiller or the gearing in a wheel steering system?

The length affects the amount the helmsman must move the tiller in order to turn the boat. It also affects the apparent 'weight' in the helm and so is obviously a critical factor. Similarly, the gearing in a wheel-steered system affects the amount the wheel must be turned relative to the movement of the rudder.

The best arrangement is a compromise between manoeuvrability and easy straight-line sailing. A long tiller or low-geared wheel is ideal for delicate upwind precision, but it makes applying large amounts of helm difficult when manoeuvring at the start, or downwind in waves. It is possible to hold a well-designed tiller at any point to change the ratio, and in the maxi class double-geared wheel systems have been tried. In general it is best to gear the system as high (heavy) as is practical.

What is the difference between under-the-stern rudders and transom-hung rudders with regard to turning efficiency?

Apart from the differences of leverage discussed earlier, another important factor affects stalling. It is often referred to as 'cavitation' but the phenomenon is actually aeration.

It is most likely to occur when the foil pierces the surface of the water. When the rudder is under high load, such as when trying to prevent a broach – in other words when you most need it – the rudder suddenly loses 'grip' and has no steering effect at all. Air is sucked down the low-pressure side of the blade allowing the water to simply break away and not create the hydrofoil effect. Foils that are fully immersed do not suffer in this way and some transom-hung rudders are fitted with 'fences' – horizontal plates at the top – which deny the air a clear passage down the foil and reduce aeration effects.

(Incidentally, cavitation is the local boiling of water on the surface of a foil, caused by the reduction in pressure created by the hydrofoil effect. It occurs on propellers, but usually only at the tips of the blades where, at high speeds, the pressure and thus the boiling point of the water is very low. Cavitation has never been known to occur on a rudder or keel.)

How can a loss of steerage be countered?

Many boats, while easy to steer most of the time, become strangely difficult to control downwind. One of the most common problems is that the rudder suddenly and completely loses grip and the boat rounds up.

In normal conditions, while the water flow around the rudder is attached and the rudder is operating as a hydrofoil, the force it creates is proportional to its angle. If it stalls, however, this instantly decreases to nearly zero and the helmsman has no control. In this instance, the aspect ratio of the rudder is the prime area of concern since the higher the aspect ratio the smaller the stall angle. In classes where some latitude is allowed in rudder shape, reducing the aspect ratio of the rudder – basically increasing the width relative to the length – will usually have a positive effect. In classes where the rudder is one-design, trimming the boat by the stern will usually help to keep the rudder immersed as the boat heels and reduce the risk of premature stalling.

Stalling is also a function of the leading-edge section shape. If a rudder stalls too easily it usually pays to make the leading edge fatter. This will increase rudder drag but the boat will be more controllable in extreme conditions.

5 Yacht rigs

Rig tune has many guises depending upon the style and complexity of the rig. In general rigs can be divided into two categories – masthead and fractional – and are further subdivided according to their complexity, as follows:

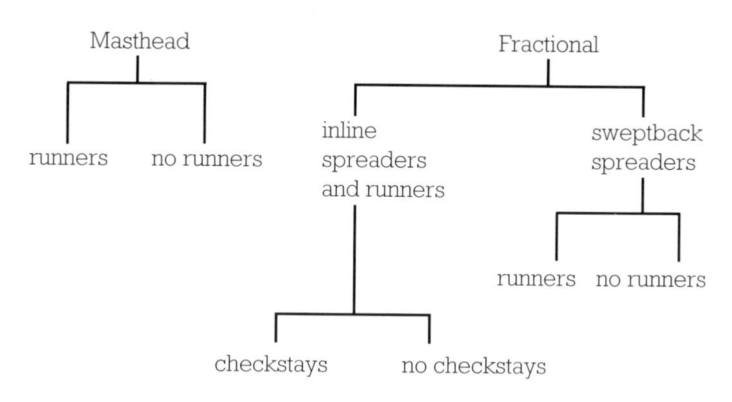

What is a mast?

A mast is a cantilevered structure which supports the sails of a yacht and holds them in their optimum shape for any given set of sailing conditions.

A mast is like the arm of a swinging crane except that the forces on it are in three dimensions rather than just two. It must be able to withstand the compression loads imparted by the standing rigging and the halyards as well as the fore-and-aft bending forces required to set the mainsail correctly.

Why does a mast need to bend?

A mast must bend in the fore and aft direction to match the set of the mainsail to a wide variety of weather conditions. Although the mainsail has its own intrinsic controls such as the halyard, outhaul, cunningham, vang and mainsheet, these controls are insufficient to give a fast mainsail shape in all windspeeds. Therefore the mast must be capable of bending to either pull fullness out of the main in heavier conditions or introduce fullness into the main in lighter conditions.

Why does a mast have spreaders?

The athwartships forces on a mast imparted by the rigging can be resolved into horizontal forces, which support the mast, and compressive forces which tend to squash it. In an ideal world a mast would be supported by stays extending horizontally to windward so that all the forces created by the stays would support it athwartships. This is obviously impossible, but the nearer to horizontal the stays, the greater their supportive forces and the less

▲ *When the mast is too straight (left) the main is too full and backwinded excessively. With the mast bent correctly (right) the draft stripes are straight and the main is 'bladed out'.*

their compressive forces. Spreaders therefore increase the angle of the stays to the mast, increasing the ratio of mast support to mast compression. Spreader lengths, although, are dictated by the angle at which the genoa has to be sheeted for upwind sailing or by the beam of the boat.

What are jumpers or diamonds?

Diamonds and jumpers are stays that support the topmast of a fractionally-rigged boat. They run over short spreaders at the hounds (the intersection of the shrouds and forestay with the mast) and are usually attached to the mast at the top and at some point below the hounds.

Diamonds are oriented athwartships and therefore affect the mast only in the athwartships direction, controlling the falloff of the mast tip. Jumpers are usually angled forwards and control athwartships bend and fore-and-aft bend in combination. On a modern mast diamonds are uncommon, since it is usually desirable to have some fore and aft control over the topmast as well as athwartships control.

Why does a mast have diamonds or jumpers?

Jumpers and diamonds enable the section of the mast above the hounds to be made considerably lighter, and also allow a degree of control and adjustment over the topmast. This has a fundamental effect on both the shape of the top of the mainsail and the leech tension in the mainsail.

There has been prolonged debate over the virtues of diamonds and jumpers, and many boats of the late 1980s sported masts without any topmast support. However, for just a small addition to the overall windage they save a great deal of weight and offer a large degree of control, so unless they are outlawed by any particular class rule, jumpers or diamonds are usually desirable.

How should jumpers or diamonds be adjusted?

Because these stays are considered part of the standing rigging of a mast, they must not be altered while racing unless this is specifically allowed. However, it is often useful to adjust them before the race starts, taking account of the prevailing conditions.

Since jumpers reduce rather than induce bend, in light airs they are often set up with more tension to maintain depth in the top section of the mainsail. This keeps the mast upright athwartships and allows leech tension to be applied via the mainsheet and vang without causing the top of the mainsail to flatten. In heavier airs the stretch in the wire or rod may allow the top of the mast to pant sufficiently to depower the rig for the conditions but if this is not the case the jumpers can be slackened off to depower the rig.

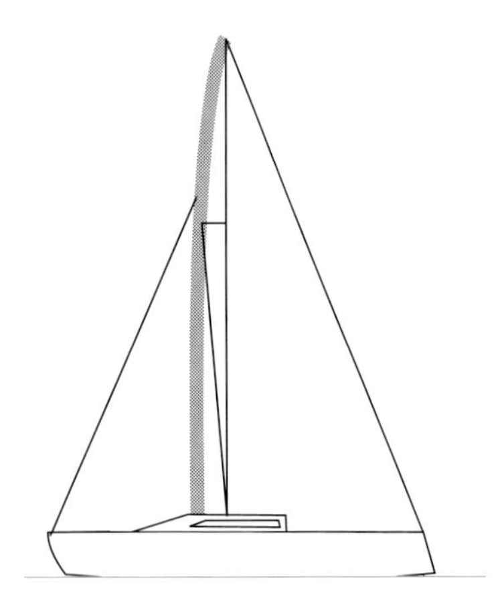

How should the rig be set up?

The tension required in the shrouds depends very much upon the nature of the rig, as follows:

Fractional rig, sweptback spreaders, topmast backstay, no runners In a rig with sweptback spreaders the shrouds not only provide athwartships support but also make some contribution to headstay tension. Therefore the amount of tension in the shrouds must vary according to the conditions. Assuming the mast is keel-stepped, the mast partners – the deck-level support of the mast – is the first point at which pre-bend (more later) is controlled. The amount of pre-bend necessary will depend upon the mainsail and the conditions, but the mast must not twist or move laterally in the gate since this will affect its bend characteristics and its ability to sustain the required rig tension.

Once the pre-bend is set into the rig by adjusting the position of the mast foot and the fore-and-aft position at the partners, the shrouds must be set to match the pre-bend and support the rig. This setting depends on the conditions.

In light airs, and up to medium airs in a seaway, it is best to induce fullness in the headsail by allowing the forestay to sag slightly. This is done by reducing the cap shroud tension. If the rig has been set up for medium airs and flat water the lowers will have been set to support some pre-bend and the act of slackening the caps will not only slacken the forestay but also straighten the

◀ *A fractional rig with in-line spreaders, topmast backstay, runners and checkstays – the Mumm 36.*

mast; so in this case it may not be necessary to adjust the lowers. Looking up the mast athwartships when sailing upwind will indicate whether the middle panels of the mast are in column. If not, the lowers may need to be tightened up.

As the wind increases the cap shrouds should be progressively wound down to increase headstay tension and keep the mast in column athwartships. Because of the pre-bend the tension in the caps will tend to push the lower sections of the mast forward into a bowed shape; the lowers control the amount of bow at the same time as keeping the mast in column.

Fractional rig, in-line spreaders, topmast backstay, runners, checkstays This is one of the most versatile and controllable rig configurations. It is the one used on virtually every custom racing boat.

In this configuration the shrouds simply support the mast athwartships, the system acting like a cantilever with the runners controlling the headstay tension and the backstay and checkstays controlling the bend. Athwartships rig tension is only necessary to keep the mast in column and as vertical in the boat as possible,

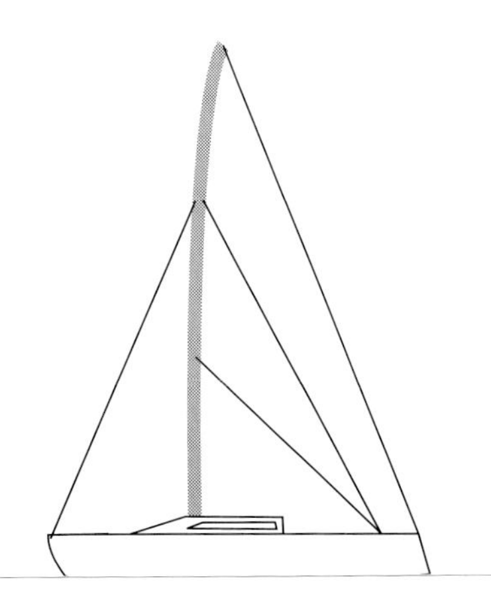

and it is not unusual, therefore, for the diagonal shrouds to be only lightly tensioned. Pre-bend is still required, for the same reasons as above, but it is only necessary in the lower panels since upper panel bend is controlled by the checkstays and the topmast backstay. Pre-bend is therefore the domain of the mast heel and partners, and again should be set up to match a given mainsail. Headstay tension is supplied by the runners.

Why is a mast usually set up with pre-bend?

A mainsail is cut with a certain amount of luff curve, and the mast must therefore be set up with pre-bend to match the luff curve of the mainsail for medium conditions. In lighter airs, where more power is required, the mast can be straightened to induce more fullness in the mainsail. In heavy airs the mast can be bent further to flatten the sail. It is critical that the mainsail and mast are matched correctly.

How much pre-bend?

The mast must be dead straight when the mainsail is as full as it is ever required to be. Pre-bend should then be introduced so that the mainsail is the correct fullness for average conditions.

On a conventional fractional rig the mast should never be pre-bent more than absolutely necessary for two reasons:

- If large amounts of pre-bend are used the mast will be less capable of remaining in column and withstanding compression loadings, and therefore rig tension. This will make it difficult to apply headstay tension because the mast will compress rather than support the headstay.
- If a lot of pre-bend is needed for an average setting of the mainsail, the mainsail is probably cut with too much luff curve and will not be capable of being either flattened or powered up properly.

What are the sectional characteristics of a mast?

The section shape of a mast is a slice through it. The section must meet several criteria:

- Low windage
- Light weight
- Required strength both fore-and-aft and athwartships.

Of course all of the above criteria conflict. For the lowest windage the mast section should be as small as possible. However, if the mast is fixed it is not possible to align an aerofoil section with the flow for all angles of sailing, so in this case a delta shape offers least resistance for the widest range of angles. A small, low-windage section may also be unnacceptably weak, since according to mechanical theory and practice the ideal combination of strength and light weight is provided by a large, thin-walled section.

The stiffness characteristics of a mast section are described by what is known as its Second Moment of Inertia (I). This is taken about any arbitrary axis – fore and aft, athwartships or even diagonally – which is known as the Neutral Axis.

Mast Sections	*Weight (kgs/mt)*	*A (mm)*	*B (mm)*	*Fore & Aft Stiffness*	*Sideways Stiffness*
2420	*0.78*	*61*	*50*	*10.0*	*7.5*
Lambda	*0.86*	*63*	*51*	*13.0*	*10.0*
C	*0.90*	*65*	*54*	*14.0*	*10.0*
Kappa	*0.91*	*67*	*55*	*16.5*	*12.0*
D	*0.97*	*73*	*57*	*19.5*	*12.0*
Stratus	*0.98*	*69*	*57*	*19.5*	*15.0*
Epsilon	*1.02*	*72*	*57*	*20..0*	*15.5*
D+	*1.03*	*73*	*57*	*19.5*	*14.0*
Nimbus	*1.16*	*68*	*54*	*20.0*	*15.5*
E	*1.17*	*70*	*54*	*19.0*	*14.0*
Cirrus	*1.2*	*75*	*65*	*28.0*	*20.0*
Gamma	*1.22*	*75*	*56*	*27.0*	*16.0*
F	*1.35*	*78*	*60*	*30.0*	*20.0*
Electron	*1.00*	*61*	*66*	*11.0*	*18.0*
Zeta	*2.20*	*85*	*65*	*31.0*	*41.0*

◀ *A range of mast sections for dinghies offered by a leading spar manufacturer. Note how the section shape affects the stiffness in each axis.*

For a yacht mast the selected Neutral Axis is usually in the fore-and-aft direction or in the athwartships direction, since the stiffness in these directions determines the ability of the mast to stand upright and not sag off to leeward, and to bend to accommodate varying mainsail shape requirements.

I relates not only to the area of material in the section but also its distribution about the chosen Neutral Axis. For a cylinder, I is the same whatever axis is chosen but in a mast, which is usually longer than it is wide, I is greater in the fore-and-aft direction than in the athwartships direction which is supported by the standing rigging. I is significant because it defines the stiffness of a mast section and therefore the ultimate bending characteristics of a mast.

The stiffness of a section is related to the fourth power of the dimensions of the section, so a mast whose diameter is doubled is some 16 times as stiff.

The same I, and therefore bending characteristics, can be achieved with either a small section with thick walls or a large section with thin walls – up to the point where the walls are so thin that they buckle under the compression loadings in the mast. Therefore the ideal section is a trade-off of size against wind resistance and disturbance of the flow over the mainsail.

What is the ideal section shape?

The ideal section shape depends on the nature of the rig. In an unstayed mast, such as a Finn dinghy mast, the section must be strong enough to support itself and bend evenly over the length of the mast. The section shape and size must therefore be tailored not only to the cut of the sail but also the weight of the sailor (a heavier sailor can harness more power and can therefore use a stiffer mast). The bend characteristics of an unstayed mast are entirely dictated by the section and the spar material.

On a stayed rig the mast is supported laterally by the shrouds and the spreaders and fore and aft by the forestay and any backstay and runner system that may be used. The section shape will depend upon the righting moment of the boat, which indicates the power of the boat and therefore the loadings a rig must withstand.

Imagine a rig attached to a model yacht and pushed over by hand. The forces in the rigging and mast will be quite small because the boat will yield to them and heel over. If the same rig is attached to solid ground the forces will begin to break the rig instead of heeling it. Similarly the loads in the rigging of a stable boat will be much greater than those in a less stable one.

There is also the question of slamming into waves which must be taken into account: as a boat slams into a wave the loads generated in the mast and rigging are momentarily multiplied by a large factor. Therefore the rig must be designed to withstand not only the loads of normal sailing but the abnormal loads generated by large seas.

The ideal section shape gives minimal athwartships bend – up to the hounds at least – with controllable, even fore-and-aft bend to match the luff curve of the mainsail. The section must also be stout enough to withstand the compression forces imparted by the tension of the rigging and the halyards.

Most metal masts are extruded from a single section and therefore have the same characteristics along the length despite the fact that the loadings vary at different places on the spar. However, there are several ways that a section can be locally modified.

- Internal or external stiffening. This is usually added to the lower panels of a mast to increase stiffness in the area of highest load without resorting to a heavier section for the entire spar.
- Modifying the section (tapering). In the top panels of most rigs some material is cut from the walls of the section to reduce its weight. In fractional rigs the reduction can be dramatic – often reducing the section diameter to much less than half – particularly if diamonds and jumpers are used.
- Chemical etching. Chemical etching is an acid-bath process by which metal can be selectively eroded from parts of a mast to alter its weight and section characteristics. It is the most accurate and therefore effective method of altering the section since, in theory, the section can be modified to the precise requirements at any position. However, it is expensive and time consuming and is usually restricted to custom racing projects.

Is aluminium the best material for a mast?

Carbon fibre is better but many times more expensive and therefore outlawed by many class rules. Carbon is better for a number of reasons:

- It is lighter for the same stiffness – or stiffer for the same weight – because carbon is stronger than aluminium. A lighter section can withstand greater compressive loads without buckling. Carbon is more brittle than aluminium, however.
- The sectional properties of a carbon mast can be tailored to suit each particular portion of the mast at the design stage by adding or omitting material. This results in a lighter, more efficient and more homogeneous mast.
- Housings for fittings such as spreader bases, sheaves and halyard exits can be engineered into a carbon mast as it is built, rather than retrofitted.

What is the significance of the shroud base?

The shroud base dictates the sheeting angle of the genoa and also the size of mast section required. On boats with non-overlapping headsails, the shroud base is usually as wide as possible – to the deck edge – because the wider the shroud base, the longer the spreaders, the greater the angle of the shrouds to the mast and therefore the greater the ratio of support to mast compression.

On boats with overlapping genoas, the designer and sailmaker should specify a minimum sheeting angle which will dictate the position of the shroud base; this will then have a cascade effect on all the factors mentioned above.

What is the difference between a keel- and a deck-stepped mast?

Deck-stepped masts are unusual on modern boats because the mast pre-bend cannot be controlled as well as that of a keel-stepped mast. The required section size is also increased.

According to beam theory the strength characteristics of a beam depend not only on the nature of the beam – in this case the mast section, the number of spreaders and the material – but also on 'end fixity'. A deck-stepped mast approximates to what is known as a hinged end while a keel-stepped mast approximates to what is known as a fixed end. According to Euler's Theory, the buckling load 'P' in an unsupported panel of a mast 'L' with a hinged end can be obtained from the following formula:

$$P = \frac{\pi^2 EI}{L^2}$$

'I' being the second moment of inertia and 'E' being Young's modulus – a measure of the strength of the material (for example, carbon is roughly four times as strong as aluminium and 20 times as strong as wood). The buckling load in an unsupported panel of a mast with a fixed end can be obtained as follows:

$$P = \frac{2\pi^2 EI}{L^2}$$

Therefore a keel-stepped mast can withstand twice the compressive stress in the bottom panel as a deck-stepped one. For this reason the bottom panels of keel-stepped masts are much longer than the upper panels where the fixity at the spreader roots approximates to a hinged end.

How much should a mast be raked?

Rake depends not upon the geometry of a rig but on the balance and performance of the boat. In earlier chapters we have seen how the helm is balanced by the relative positions of the Centre of Lateral Resistance (CLR) and the Centre of Effort (CoE) of the rig. To maintain the required small amount of weather helm it is necessary for the CoE to be slightly aft of the CLR; about 12 per cent of the waterline length is usual.

The position of the CoE is determined by the position of the mast in the boat. The mast can be moved forward and aft at the foot and at the base, and it can be raked forward and aft. If there is insufficient weather helm (about five degrees in 10 knots true when sailing upwind) or if the boat appears slow upwind compared to its rivals, the mast can be moved bodily backwards – by moving it at the heel, the partners and by lengthening the forestay – or it can be moved independently at any of the above points.

Increasing the forestay length alone will increase the pre-bend because, with the mast fixed at foot and deck, when the hounds move aft the spar has to bend. Increasing the forestay length and

moving the heel forward will maintain the same pre-bend; increasing the forestay length and moving the partners back will reduce the pre-bend. Which of these is appropriate will depend on where the mast is situated in the gate and at the heel, and whether the pre-bend is judged correct before the mast is moved.

Aft rake has a negative effect downwind because it holds the spinnaker closer to the boat and spoils the flow of air over the mainsail. If the mast is raked forward downwind the air hitting the mainsail escapes over the top as well as off the sides which seems to be faster.

Therefore it is advisable not to rake the mast further aft than is necessary because although it will be possible to remove some of the rake downwind, forward rake is limited by the point at which the mast inverts. Inversion is when the mast, instead of either being straight or bent with the middle forwards and the tip aft, is bent with the mid panels aft and the tip forward. Because the mast and all its fittings are designed to bend aft they are likely to fail if this happens, so mast inversion should be avoided at all costs.

▲ *Allow the mast to be pushed upright on the run to improve the airflow over the mainsail (right) and move the spinnaker away from the boat. If necessary take a halyard to the headsail tack fitting and pull it forward as far as possible without inverting the mast.*

Should a mast bend athwartships?

In an ideal world a mast should stand exactly straight athwartships with controlled tip fall-off to depower the mainsail in stronger winds. However, despite the use of rods made of exotic materials such as carbon and cobalt, it is impossible to eliminate rigging stretch entirely and keep the mast exactly straight. Therefore the optimum condition is with the mast falling off as little as possible, but evenly along its length. In custom racing boats the fall-off is minimised by powerful hydraulic mast jacks that lift the base of the mast and induce massive amounts of rig tension.

6 Dinghy rigs

A dinghy rig is not dissimilar to a keelboat rig with sweptback spreaders, and the main criteria for a dinghy rig certainly differ very little from those of a keelboat rig: to support the sails in the most efficient manner, in the widest possible range of conditions, while remaining as light as possible.

The most significant difference is that whereas the righting moment of a keelboat is usually fixed by the designer – with some variation owing to the weight of the crew – the righting moment of a dinghy is solely dependent upon crew weight and varies as their weights vary.

In theory a heavier crew exerts a greater righting moment which requires a heavier mast section and stronger rigging to keep it in the boat. In practice, however, crew weights in a class vary little since the optimum all-up weight for a crew soon becomes apparent for the best all-round speed. The optimum weight becomes an important reason for choosing a particular class; crews that are heavier and lighter than the optimum usually find themselves uncompetitive and attracted to another class.

What stays are used on a dinghy rig?

Dinghy rigs are usually much simpler than keelboat rigs, apart from complicated boats like the 18-Foot Skiff, International 14 and Ultra 30. Almost without exception they feature sweptback spreaders and no backstay, with one set of cap shrouds passing over the spreaders and terminating at the hounds, and a forestay or wire jib luff. The stays on some dinghies are low-stretch rods but in most cases they are wire.

Why so simple?

In general, dinghy rigs are simple because it is impossible to alter and control a complicated rig with a crew of just two. Because of this dinghy rigs are usually more self-regulating than the more complicated keelboat rigs. It is rare, for instance, for dinghies to use jumpers. The upper section of the mast is tailored to the optimum crew weight, and the top bends to allow the leech to open as the wind increases without the crew having to alter the settings while underway. An exception is the Star, where controlling the mast bend with runners and keeping the spindly rig in the boat is intrinsic to sailing in the class successfully. In general, however, runners and stays that must be altered simply in order to tack or gybe are not used in dinghies.

Backstays are also usually impractical because there is simply not enough boat aft of the mainboom to allow a backstay to be fitted without severely limiting the amount of roach that can be carried in the mainsail. So there is usually little choice but to use a sweptback spreader configuration, and the single spreader rig, being the most simple and versatile, is usually favoured.

6003
K·62
6515
6954

Dinghy rigs are not without complications of their own and these are to be found in the combination of rams, screws and struts which control the lower sections and by various systems of shroud and forestay adjusters which allow alterations in rig tension while racing.

What is the significance of mast section?

In some ways the mast section on a dinghy is more significant than on a keelboat where the greater number of adjustable stays provide greater inherent versatility. The controls for a dinghy mast are usually fairly simple and the versatility is provided by the mast itself, particularly above the hounds where mainsail leech control is almost exclusively a function of topmast bend.

The size of the section will depend upon the optimum all-up crew weight of the particular class. It also depends upon whether trapezes are to be used because crews on trapezes increase compression in the mast.

Some years ago it was fashionable to use small, heavy mast sections, the reason being that the smaller section affected the airflow over the sails less. Modern mast sections are more sophisticated, and the larger, lighter sections of today offer less disturbance than their smaller predecessors.

On most dinghy rigs the section shape and size varies at different heights. Below the hounds the mast must be as straight as possible athwartships and must be able to bend fore and aft evenly to match the luff curve of the mainsail. Above the hounds, where the topmast has an influence over sail shape but is not controlled by stays, the section shape is more critical: as the topmast flexes it tends to open the leech of the mainsail and depowers the sail.

The point at which this depowering occurs is important because if it depowers when the wind is too light the boat will not point and will not sail well upwind. If it does not depower the sail early enough, however, the boat will be overpowered and the helmsman will be forced to sail with the traveller further down which will close the slot.

The optimum topmast section will vary from crew to crew: a lighter crew will require a more flexible topmast than a heavy crew. However, it is critical that the section does not depower the mainsail too early in medium conditions so it's worth taking care when choosing sections.

In practice, the best section for a particular class is usually the section already being employed by the mast manufacturer, who will have developed the section to suit the boat and its optimum crew weight. At the top level it might be worth experimenting with mast modifications, but in general it pays to use the generally accepted mast for the boat and concentrate on using it well.

◀◀ The slim, whippy mast of the Star is supported by a complex rig with runners, but most dinghies have simple rigs that need much less attention on the water.

How do spreader lengths affect section size?

Spreader lengths have a significant effect because they alter the angle the shroud makes with the mast at the hounds. The greater the angle, the less the compressive forces in the mast for the same sideways supporting force. Therefore a boat with short spreaders requires more rig tension to support the mast than a boat with long spreaders.

In boats with overlapping headsails the spreader length affects the sheeting angle of the headsail and so a compromise between spreader efficiency and sheeting angle must be reached. In boats with non-overlapping headsails the only constraints are those of windage and practicality. It is therefore ideal to fit spreaders that are long enough to keep the mast in column at all times using practical levels of rig tension.

How stiff should a mast be?

A dinghy mast should always be as stiff as possible for the lightest weight and smallest section. As with a keelboat the three considerations are in conflict but in general the stiffer the mast the greater the rig tension it will be able to support, the straighter the forestay will be and the faster the boat will be upwind.

What gadgets are best used to control mast bend?

There are many gadgets available to control mast bend, particularly in the lower sections. Some boats are fitted with adjustable mast steps which allow the pre-bend to be altered while under way. More common is a device such as a screw or a ram which controls pre-bend by moving the mast fore and aft at deck level. These devices are easy to adjust and calibrate, and are generally quite light and compact.

However, they are less effective than a mast strut, which operates higher up the mast, nearer the point of maximum bend, and exerts more control over the mast, both in terms of accuracy and evenness. A strut also helps to oppose the forces induced in the mast by the vang; this makes vang-sheeting a more efficient option while still maintaining precise control over mast bend.

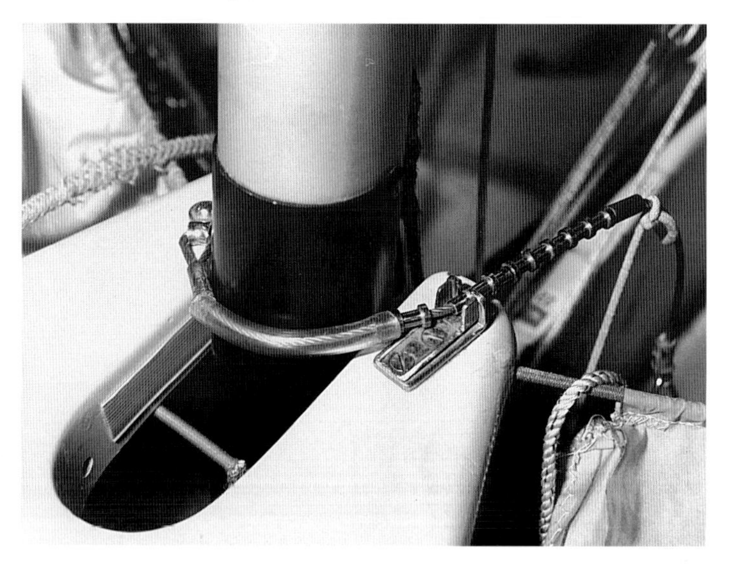

▶ *Altering the mast position at deck level controls the pre-bend. This simple wire and ferrule system is effective, but is difficult to adjust underway.*

Should dinghy masts be pre-bent?

Because, on the whole, it is possible to adjust the lowdown bend of a dinghy mast while sailing through the use of struts, screws and adjustable shrouds, pre-bend is not necessary. A mast can be bent at any time to match the prevailing conditions. However, where it is not possible to move either the partners or the heel, some pre-bend will be necessary as described in the section on keelboat masts. This pre-bend should, where possible, be adjusted on a race-by-race basis.

On boats such as the 18-Foot Skiff and International 14 it is necessary to pre-bend the mast over its entire length to enable it to withstand the high forward loads imposed by masthead spinnakers and gennakers without inverting.

How and why is rake altered?

It usually pays to rake a mast aft for upwind sailing and forward for downwind sailing, the degree of rake increasing with the wind strength.

There are many theories as to why it pays to rake aft upwind, but none are conclusive. Perhaps the most compelling argument is that the centre of effort of the sail plan moves forward as the wind increases: the mainsail is increasingly depowered and downtracked and contributes more to the depowering than the headsail.

Conversely, downwind it pays to rake forward, the most plausible explanation being that it helps to move the spinnaker away from the boat and alter the airflow over the mainsail so that the air is escaping from all its edges rather than just horizontally around the leech. Raking forward also moves the centre of effort of the rig forwards, which increases the effect of the rig 'dragging' the boat through the water rather than trying to 'push' it. This means that less helm is necessary, and the boat naturally tracks straighter.

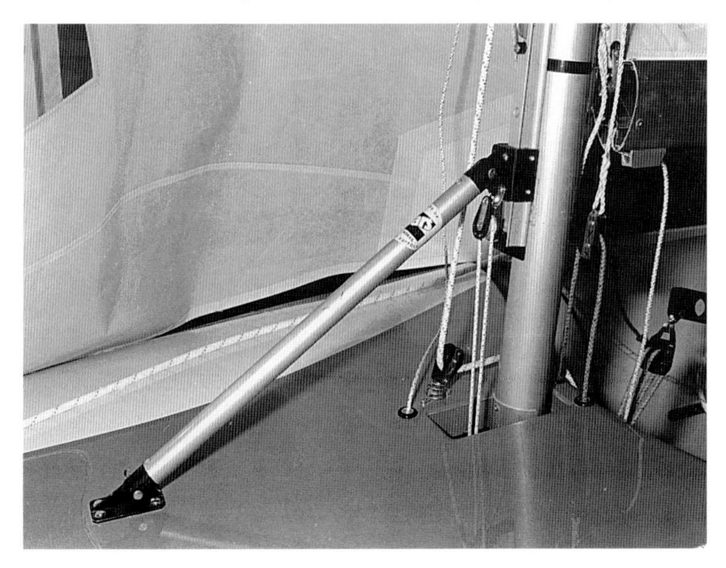

◄ *A mast strut gives precise, quick and effective control over mast bend in the lower sections.*

On most keelboats the positions of the mast foot and partners are fixed, with the mast set up for upwind sailing. Therefore the amount the mast can be raked forward depends upon the amount of inversion the section will tolerate. A dinghy is more flexible because the jib luff usually replaces the forestay or takes the principal sailing loads. It can therefore be adjusted to allow the mast to rake aft or to pull it forward. Also, by using a mast ram or strut it is possible to provide movement at deck level which means that the mast can be raked further forward without inversion. And finally, where adjustable shrouds are used, the combination of jib halyard, mast gate and shrouds allows the mast to be raked to virtually any desired position.

How these adjustments are made is a matter of both technology and fashion. In the past screws and deck level rams have been popular, but they have generally been superseded by struts. Jib halyards are usually moved by muscleboxes, as are struts, while shrouds are either hydraulic or musclebox powered. The criteria are simplicity, ease of operation, efficiency and low weight, and equipment manufacturers are continually developing new systems that improve one or more of these factors.

How much rake?

In certain classes like the Flying Dutchman and the 505 it has been found that a huge amount of aft rake is fast, particularly in a breeze. The question of how much rake depends on how the boat was designed in the first place, particularly with regard to the fore-and-aft position of the mast relative to the keel and rudder.

Contrary to popular belief, many yachts and boats are curiously designed hydrodynamically and have their own peculiar requirements when setting up. The keel on the J/24, for instance, is set a long way aft and so it is very hard to develop weather helm on the boat upwind. Therefore the forestay on a J/24 is always set at maximum to rake the mast as far aft as possible, thereby moving the centre of effort back to create a reasonable degree of weather helm.

Generally speaking upwind the mast should be raked far enough aft to create three to five degrees of weather helm, and the amount of rake required to achieve this will vary from boat to boat. It is important to realise that the helm angle is more important than the weight in the helm. This actually applies more to keelboats than dinghies since the loads on a keelboat rudder are generally higher. The principle is that the balance of the rudder (the position of its pivot point relative to its centre of effort) has an effect on the weight of the helm, as does the length of the tiller (or the gearing ratio in a wheel system). Neither of these two factors has a direct bearing on the amount of lift a rudder is creating.

Whether the actual amount of weather helm is three or five degrees depends upon what is fast in a particular class: if a boat has a large centreboard the rudder is superfluous to the sideforce requirements. In such a boat less weather helm means not just less lift from the rudder but less drag, which is faster. Conversely, if the centreboard is small the boat will require more weather helm to ensure that the rudder contributes sufficiently to the overall lift force. The best way to assess a particular boat is either to gauge its performance against other boats or to monitor the amount of leeway the boat is making, which should be no more than about seven or eight degrees at worst.

Downwind, when reaching, it is best to balance the boat so that the helm is neutral and the rudder is not creating lift, so minimising the drag. Lifting the centreboard and raking the mast upright helps achieve this, as does heeling the boat. When running, the mast should be raked as far forward as is practical.

How much should the mast be bent?

The general rule is that the rig is most efficient when the mast is straight. However, some bend is necessary to adapt a mast to a particular mainsail. This is covered in the keelboat mast section.

What effect does spreader angle have on mast bend?

If the spreaders deflect the shroud forwards, increasing shroud tension under normal sailing loads will tend to push the spreader aft and straighten the mast, effectively making it stiffer. If the spreader deflects the shroud aft, increasing shroud tension will tend to push the mast forwards and bend the mast. Therefore the spreaders should be adjusted to suit the crew weight and mainsail shape. Using the same mainsail, a lighter crew will need to flatten the sail earlier up the wind range, and the spreaders will need to be angled farther aft than for a heavier crew. Similarly, a relatively flat mainsail will require spreaders set further forward to limit mast bend.

Once again, because it is generally not possible to alter spreader length and angle while sailing, the spreaders must be set so that the rig is self-regulating as the wind varies. They must deflect the shrouds such that in light airs, when rig loads are light, they straighten the mast and increase the fullness and power of the mainsail. As the wind increases and the rig loads increase, they should bend the mast to flatten and depower the mainsail.

505
GBR8458

7 Sails

What is a sail?

A sail is a wing-like mechanism which, like a keel, divides steadily flowing air down each of its two sides, causing the air down the leeward side to travel further and faster. This creates a low-pressure and a high-pressure side, and a resultant force which can be split into two components: driving force and side force.

Why are sails triangular?

The most efficient shape for a sail is not triangular, but elliptical. Unfortunately an elliptical soft sail is difficult to design, expensive to build and prone to wearing out quickly. Therefore most class rules limit the girth length at various points up the height of the sail which results in a more triangular profile, the idea being to control costs and ensure that sails have a reasonably long working life.

What is the difference between various aerofoils?

For any given set of stable conditions – constant airflow, constant angle of attack and no pitching or rolling – there is an optimum aerofoil shape that will provide maximum lift for minimum drag. As with a keel, the lift/drag ratio is the measure of the efficiency of a sail's foil section. Again, unlike an aeroplane's wing, the sail must operate at many angles and in many wind speeds and sea conditions, so different aerofoil shapes need to be induced into a sail at different times – fat ones, thin ones, fine-entry ones, blunt-entry ones – and each different aerofoil shape has its own specific application.

On a boat there are so many variables that every foil – keel, rudder or sail – has to be a compromise, but a soft sail is quite versatile because its material can be stretched to flatten it, move the position of flow and alter its camber characteristics. Sailmakers are constantly striving to develop materials and panel layouts that exploit this advantage.

One of the biggest problems is that on a conventional boat every foil must operate on both tacks and must therefore be either symmetrical – as keels and soft sails are – or be capable being altered to take up a particular section on each tack by the use of adjustable wing masts or keels with trim tabs.

In general terms, a foil with a blunt leading edge will operate at a reasonable level of efficiency over a wider range of angles of attack than a thin foil; as the angle of attack increases, the thin foil will stall before the thick foil.

Is camber the same as thickness?

In a symmetrical section yes, but in a non symmetrical section no: the thickness of the foil is its physical width whereas its camber is the distance between a straight centreline and the maximum thickness of the foil.

What is the difference between an aeroplane wing and a sail?

The principal difference is that a wing need only provide lift in one direction (upwards) while a sail must provide lift equally in two directions (port and starboard). An aeroplane wing also operates at a steady cruising speed for most of its working life.

Why are sails flexible?

Some are not, but these are usually for specialist applications. They are heavy, complicated, expensive and not usually versatile enough for general use, although in their specific applications they are much more efficient than soft sails. Soft sails are more practical, they can cope with a wide range of conditions, can be reefed and can be taken down and stored when necessary.

What is the ideal profile?

The ideal shape varies a great deal depending upon wind strength, sea state and type of boat. In the absence of class rules a tall narrow profile with an elliptical top is most desirable. The tall narrow profile ensures a high aspect ratio, keeping tip losses to a minimum and ensuring they affect only a small percentage of the sail area. The elliptical top minimises the size of the tip vortex and hence reduces one element of drag on the rig. however, like a deep narrow rudder, tall narrow rigs are more prone to stalling.

What is lift/drag ratio?

The lift/drag ratio is the relationship between the driving force created by a sail and the sideforce that must be opposed by the keel. The higher the lift/drag ratio the more efficient the sail.

What is the ideal section?

The section shape is the shape of a cross-section cut through the sail horizontally (in line with the airflow). Because a sail is flexible it can be adjusted using the various sail controls to take up a number of different shapes. These controls all affect the camber of the sail: the ratio of the depth to the chord length (camber), the position of the maximum depth (draft position) and the angle of a particular section of the sail relative to the other sections and the wind (twist).

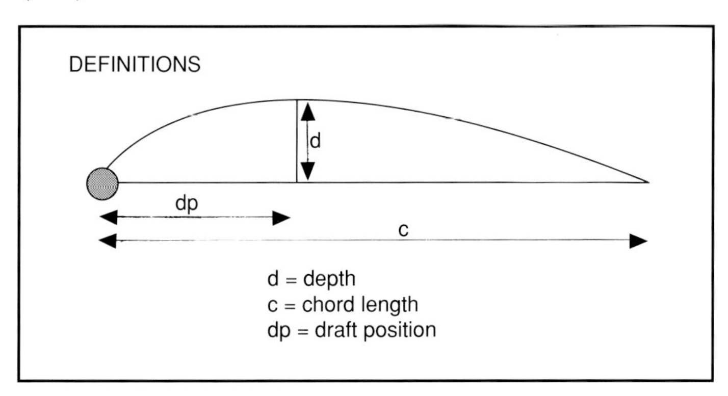

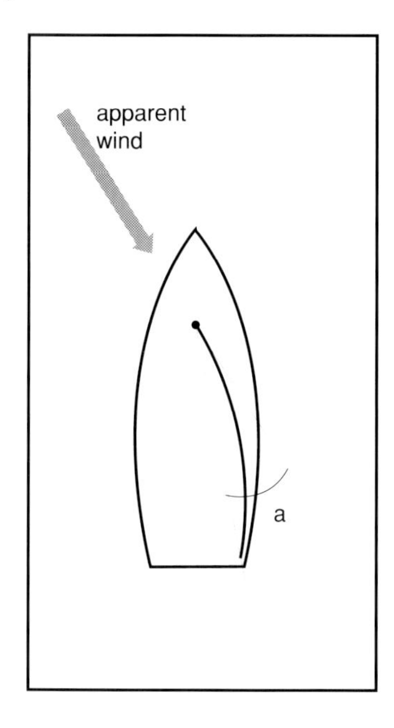

▲ *The angle a is the angle of attack*

There are two principal types of sail sections, each of use in different conditions. A round-backed section is a powerful sail with high lift, high drag properties. It is also a section that helps high pointing, especially in a mainsail. However, the danger with a round-backed sail is that although it is powerful it is prone to creating high drag forces and, if oversheeted with the leech hooked, it will prove to be very slow. A round-backed sail is most effective in light and moderate breezes when high pointing and high power are required.

In contrast, a straight-backed sail has low lift but low drag. When the sail is used at the top end of the wind range it is useful for the leech to be straight to reduce power and exhaust the wind off the sail. The problem with this type of sail – particularly a mainsail – is that when it is twisted off to depower the twist is difficult to control; there is a danger that the whole back section of the sail will flap open to the detriment of power and pointing.

What is angle of attack?

The angle of attack is the alignment of a foil relative to the air flowing over it. The centreline of a solid foil is defined as the line which, when aligned with the flow, causes the foil to produce no lift. On a flexible sail the centreline is usually taken to be the angle of the boom.

▶ *The mainsail trimmer plays the sail using the mainsheet fine tune (left hand) and the traveller (right hand).*

◄ ▲ *Good genoa trimming technique: trimming from the weather side keeps the crew weight up.*

What is stalling?

A sail stalls when the angle of attack is so great that the air flows past the back surface of the sail instead of flowing around it. This severely impairs the lift, even though the air continues to flow around the front surface reasonably cleanly.

What is camber?

The camber in a sail is the ratio of the maximum depth to the chord length at any point.

What is twist?

Twist describes how the angle of attack changes in various sections of a sail. The angle of the boom dictates the angle of attack of the lower part of the sail, but because the sail is flexible the boom angle of attack cannot be maintained all the way up the sail and increases towards the head. The flatter the head of a sail, the more difficult it becomes to control twist. Twist control is the limiting factor affecting the head profile of a sail.

What is aspect ratio?

Aspect ratio is the relationship between the luff length of a sail and its area. For the same area two sails could have vastly differing shapes: one might have a long foot length and a shorter luff length, another might have a shorter foot length and a longer luff length. The latter sail would have a higher aspect ratio.

In theory high aspect ratio sails are more efficient for upwind sailing but there are several practical limitations, as follows:

- The higher the aspect ratio the narrower the chord of the sail and therefore the greater the influence of the airflow disturbance caused by the mast.

▶ *Twist is critical in headsails when two-sail reaching. It is possible to eliminate twist by moving the snatch block forward along the rail, but this makes the base of the sail round up which, if carried too far, is slow. When reaching with a headsail look for a compromise between overtwisting the top of the leech and over-rounding the foot. Owing to its high aspect ratio this No. 3 sets with little twist. In this shot the base is too rounded.*

- For offwind sailing a high aspect ratio sail is not as efficient.
- In waves a high aspect ratio sail is more 'twitchy' and prone to stalling.
- The centre of effort, being higher, causes a greater heeling force for a given driving force.

What is true wind?

The true wind speed is the speed of the wind relative to the land. In waters where no tide or current is present, this is also the speed of the wind over the water. In tidal conditions, however, the modified wind – the wind relative to the moving body of water – is also significant. Because the speed of a boat over the sea bed will vary from tack to tack or gybe to gybe in tidal conditions, the modified wind will also vary; this will affect how the sails must be set and trimmed.

The true wind direction is the direction of the wind relative to the land, while the modified wind angle is the angle of the wind relative to a moving body of water.

What is apparent wind?

The apparent wind speed is the wind speed as experienced on board the boat. It is affected by the speed and direction of travel of the boat through the water. Similarly the apparent wind angle is the angle of the wind as experienced on board the boat, and this is also affected by the speed and direction of travel.

Which is the most significant?

In terms of actually sailing quickly, the apparent wind speed and angle are most important. The apparent wind angle affects the setting and trimming of the sails, and the ratio of apparent wind speed to boatspeed is a measure of sailing efficiency. However, the true wind speed and direction are of fundamental importance for selecting sails for future legs of the course, setting up from tack to tack or detecting windshifts. Knowing the true wind speed enables you to calculate modified wind speed and direction trends, and calculate apparent wind speeds and angles for future legs. It is relatively easy to calculate the modified wind speed and angle using a basic set of yacht instruments, but determining the true wind speed and direction requires accurate data on the strength and direction of the tide or current.

▲ *Adjusting the genoa car to control twist.*

THE MAINSAIL

What is a mainsail?

A mainsail is a sail attached to the main mast, as opposed to a sail set from a stay or wire luff (headsail) or free-flying (spinnaker or gennaker). A mainsail may be set alone (una-rig) or behind a headsail. This affects the shape of the mainsail because when a mainsail is set behind a headsail the airflow over the sail is modified and this has to be allowed for in its cut.

What are the peculiarities of the mainsail?

Because a mainsail is always set whether sailing upwind or downwind, it must be versatile. Upwind the sail acts like an aerofoil, with the air flowing from leading edge to trailing edge. When running dead downwind the sail is simply acting as a 'dam' for which the main criterion is not its shape but its area, and maximising the projected area when running is the main consideration.

Why is a mainsail attached to the mast?

Because a sail creates high and low pressure sides any holes in the sail, particularly towards the leading edge, allow air to escape from the zone of high pressure to the zone of low pressure. Therefore a sail which runs in a luff groove is more efficient than a sail set on slides. However, practical considerations often dictate that a mainsail is set on slides, particularly a large fully-battened sail.

How and why is a mainsail matched to a mast?

Since a mainsail is set on all points of sailing and at all wind speeds it has to be set in a wide variety of shapes. This is why masts are flexible – the mast takes shape, or fullness, out of the sail when bent and puts shape in when straightened. A mast will have bend characteristics designed into it by the mast designer and can be adjusted by the use of backstays, runners and checkstays (or by the shrouds if sweptback spreaders are used). A mainsail is usually cut with a certain amount of 'luff curve' (roundness in the forward edge of the sail), and this means that the mast has to be set up for average conditions with some pre-bend. So when the mast

▼ *On the left the main is over-twisted, as the boat is too upright; she is losing both power and pointing ability. On the right the twist is correct.*

is bent further the sail is flattened, and when it is straightened the sail is made fuller. It is important that the luff curve matches the natural bending characteristics of the mast, otherwise when the mast is bent or straightened the sail will not alter shape uniformly from foot to head.

How much luff curve is recommended?

Although some luff curve is necessary for a versatile sail, the amount varies from class to class and mast to mast: excessive luff curve can have detrimental effects. Firstly, in classes where sail area and girth length are restricted, any area that is built into the luff of the sail has to be traded against sail area in the leech, so more luff curve means less roach and therefore less effective area. Secondly, large amounts of mast pre-bend will reduce the stability of the mast, the compression loading it can withstand, and therefore the amount of headstay tension that can be achieved. So the smaller the luff curve the better, with the proviso that the sail must be able to achieve maximum power (a camber/chord ratio of about 15 per cent) with the mast dead straight.

How should a mainsail be cambered?

The ideal amount of camber varies with height, because since a soft sail is always prone to twist the higher sections are operating at a greater angle of attack to the wind than the lower sections. Reducing the camber in the upper sections of the sail effectively reduces the angle of attack, and therefore helps to reduce effective twist. Mainsails are usually cambered about 15 per cent in the lower sections, and this is reduced to around eight per cent towards the head, assuming zero wind shear.

In practice there is often a difference between the wind speed at deck level and the wind speed at the masthead owing to friction affecting the wind passing over the water (much the same as in the boundary layer where water passes over the hull). The difference in wind speed causes a variation in apparent wind angle between the deck and the masthead. This in turn requires that the sail is twisted to accommodate the prevailing apparent wind angle (see section on twist).

How does the mainsail affect pointing ability?

The mainsail has a fundamental effect on the boat's pointing ability. The leech area of the sail is particularly critical because it provides what helmsmen call 'bite'. This is because the centre of effort of the genoa is forward of the centre of lateral area (CLR) and the centre of effort of the mainsail is behind the CLR. So if the genoa is sheeted and driving fully and the mainsail is flogging the boat will tend to bear away – and vice versa if the mainsail is sheeted and the genoa is flogging.

It is therefore essential to maintain power in the leech of the mainsail to produce a degree of weather helm and keep the boat tending to luff towards the wind. However, if the mainsail has too much leech tension or is oversheeted towards the centreline the

▼ *Feel in the helm is created by mainsail leech tension. An open leech reduces weather helm and adversely affects pointing. A closed leech helps pointing but can slow the boat if overdone.*

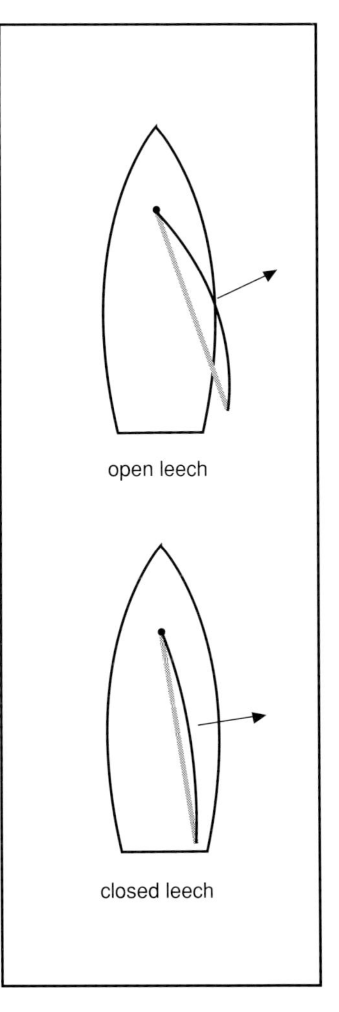

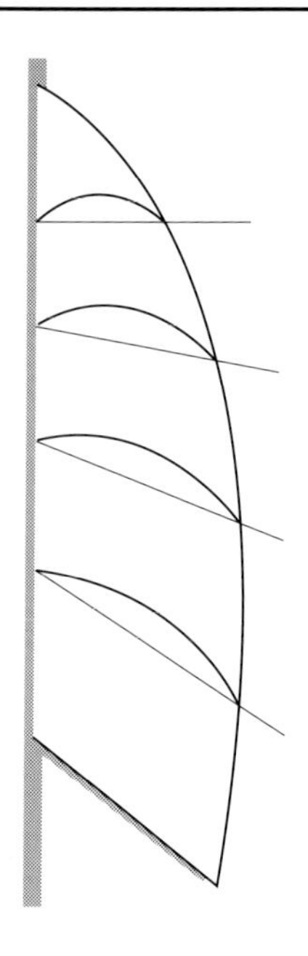

▲ *In this view from above and astern, the sections up the mainsail are increasingly twisted relative to the boom angle.*

boat will heel excessively and much of the resultant force of the mainsail will be resolved into sideforce rather than driving force.

Why must mainsails be twisted and when?

In many ways the characteristics required of a mainsail are continually in conflict. In order to maintain pointing ability it is necessary to reduce twist, but in order to reduce the heeling force it is useful to twist the top of a sail. To compound the problem, when sailing in waves a whole different set of criteria are introduced.

Ideally, a mainsail should be set so that each section stalls at the same moment. We usually place telltales up the leech of the mainsail to allow us to see when stalling is taking place. However, because at each vertical point in the sail the chord length and the camber are different, it is necessary to vary the angle of attack at different heights.

In moderate winds and flat water the mainsail is working in relatively stable conditions, and will be less prone to stalling than, for instance, in light winds and lumpy waves. Therefore the sail needs less twist, and a flatter leading edge – which is more sensitive to angle of attack – can be accommodated.

In light airs, when the mainsail is fuller and therefore more powered up, the top of the sail must be twisted more to maintain the angle of attack relationship between the shallower top sections and the lower sections. Therefore the amount of twist required in a sail is related to the camber of the sail for a given set of conditions.

What is wind sheer?

Wind sheer is the variation of wind speed with height above the water. As air travels over water there is a frictional force between the air and the water, so the air at 100 feet above water level is travelling faster than the air one foot above the water. The difference in velocity, or the velocity profile, depends upon the temperature difference between the air and the water, the humidity of the air and the temperature of the air, so there is often more wind sheer in the Mediterranean, say, than in the English Channel.

Owing to wind sheer the wind velocity varies slightly over the height of the rig, and therefore so does the apparent wind angle (because apparent wind angle is related to true wind speed). This obviously affects the amount of twist needed in the sail to maintain the desired angle of attack. It is therefore important to check the wind sheer by observing the relative stalling points of the telltales on the mainsail or genoa, and adjusting the sail settings to suit. For example, if the top telltales are lifting and the lower ones streaming when the sail is set normally this indicates wind sheer, and the top of the sail must be sheeted in relative to the bottom section.

It is said that wind sheer affects the true wind direction at different heights. This is an effect that usually occurs only near the coast where the wind is bending (refracting like light through a prism) as it crosses the land/sea interface. Such a situation makes different settings for different tacks necessary, again using the telltales to indicate the correct trim.

What do telltales show and where should they be positioned?

Telltales indicate the nature of the airflow over a sail. In general the air flows horizontally across the sail, although as the boat heels the air tends to flow slightly upwards as it does when escaping out of the top of a twisted spinnaker or genoa. When they are streaming the telltales show whether the air is flowing horizontally or slightly upwards, and if they are circling they show that the sail has stalled at that particular point.

On a mainsail the most critical part of the sail is the leech, which should have telltales at regular intervals. When the telltales are streaming evenly off the leech they indicate that the air is flowing cleanly around each side of the sail. When they start to break it is an indication that the sail is either oversheeted or undersheeted. If the sail is oversheeted the telltale will disappear around the back of the sail, while if it is undersheeted it will circle.

It is also useful to attach a row of telltales horizontally near the top of the sail to indicate how the air is behaving up there: the top sections, being short chord lengths, are more twitchy and more prone to stalling than the lower sections.

▲ *The battens support the roach of the sail – the convex area behind a straight line drawn between the head and the clew of the sail.*

Why have battens?

On a racing mainsail the line between the head and the clew is not straight but convex, and the area behind a straight line drawn between the head and the clew known as the 'roach'. Since the sail is made of flexible material the roach needs support to stop it simply flapping off to leeward. The battens provide this support.

The batten length may be unrestricted or limited by a particular class rule. In the absence of restrictions it is usually more efficient to use full-length battens which give the sail a semi-rigid aerofoil shape. However, fully-battened rigs are less versatile because they can conform to only a narrow range of shapes. Regardless of length the battens should be carefully tapered to suit the sail, both in stiffness and the nature of the curve, particularly at the top of the sail which is highly sensitive to batten influence.

How do battens control the shape?

There are two ways: the shape of the batten and the tension with which it is inserted into the sail.

Battens should be stiff enough to support the roach of the sail, even when high sheet and vang loads are used, and if a batten is too flexible a crease will develop from the head to the clew of the sail. However, as each batten extends into the body of the sail it should be tapered to take up the camber shape of the sail. The most common problem area is the top of the sail because the top batten extends across a high percentage of the sail chord.

How do mainsail controls affect the mainsail shape?

Of all the sails, the mainsail is the easiest to control. The sail is attached to the mast over the entire length of the luff and is controlled along the foot by the boom. The bend of the mast affects the camber or fullness of the sail, as does the outhaul tension along the boom. The tension in the halyard and cunningham affect the nature of the camber or the position of the fullness. The tension in the vang and the tension in the mainsheet affect the relative angles between different horizontal sections and the position of the traveller affects the general angle of the sail to the wind.

How should the fullness of the mainsail vary?

In general a mainsail should be flat in very light airs in flat water to give the air an easy path over the sail without needing to bend too much. It should be deepened to its fullest in light airs and gradually flattened as the wind increases until, when the boat is overpowered, the sail is 'bladed out', or flattened. Waves have a considerable effect on the fullness needed in a mainsail but, in general, the fullness should be a maximum of 15 per cent of the chord length in light airs and be progressively flattened as the wind increases.

How and where should the maximum depth be positioned?

Unlike a genoa, in which the position of the flow is varied according to the conditions, the position of the flow in a mainsail should be maintained at about 50 per cent of the chord length aft of the mast. If the sail is cut and matched to the mast correctly this will mean that in light airs the halyard is not taut and there is no cunningham tension. As the wind increases the halyard is gradually brought up to maximum tension and after that the cunningham is used to maintain the flow position.

The flow of the mainsail is critical because if it is too far forward not only will the sail be likely to stall but the aft section of the sail will be too straight; the sail section will therefore lack power and will not provide even lift over its chord length. This is known as 'flat-backed'. Conversely if the flow is too far aft the leech of the sail will act as a scoop and the sail will generate more sideforce than driving force. This is known as 'round-backed'.

How much twist should be used at any time?

The amount of twist required depends on the specific conditions and is really a matter of feel. In a vang sheeting mainsail system, twist is controlled entirely by the vang with the mainsheet and traveller simply controlling the general angle of the sail to the wind. In a system where leech tension cannot be provided by the vang alone, the vang, mainsheet and traveller are used together to provide the desired combination of leech tension and angle of attack.

At any particular vertical position in the sail there is an optimum attack angle such that the section is 'eased' until it is on the point of stalling. At this point the lift/drag ratio of the section is at maximum. So in theory the maximum drive from the sail will be obtained when all the sections are on the point of stalling at the same instant. However, because the section length and shape at each height is different on a triangular sail, the angle at which each section stalls

is different. To compound the problem, the air near the top of the sail tends to flow upwards to escape over the edge at a narrower point of the sail. Consequently the top sections of a sail are not operating as efficiently as the lower sections although, because of the effects of sheer, they are operating in higher wind speeds.

Because each section of the sail must be at the point of stall for maximum efficiency, the angle of each point of the sail to the true wind depends upon two things: the nature of each point of the sail – its section shape and amount of camber – and the apparent wind speed and direction it is experiencing. So if the top part of a sail is flatter than the lower part, its angle of attack for a given wind speed should be more than that of a lower section.

However, because of wind sheer the apparent wind speed is greater at the top of the mast than at the bottom, and the apparent wind angle is therefore further forward. Since the profile of the wind sheer is generally held to be greater in lighter conditions, in theory the sail should be twisted less. However, the practical effect of less twist in light airs is to stall the upper sections, since they are more sensitive because of their short chord length. Because of this the mainsail is generally over-twisted in light airs to prevent stalling. The telltales can be used to detect when the sail is at the point of stalling – indeed setting the sail for the prevailing conditions will depend largely upon the behaviour of the telltales.

What effect does the material have on a sail's qualities?

The material has a fundamental effect on a sail's qualities in two areas. Firstly weight: the lighter a sail, the lower the centre of gravity of the whole vessel; hence the higher the stability and the lower the pitching forces. Secondly shape: modern controllably-bending masts must be matched to sails that are capable of adapting their shape to the mast bend.

Modern sail materials can be broadly split into two categories: films and woven cloths. Films, such as Mylar, have consistent stretch properties in any direction; woven cloths have different stretch properties along each of the thread lines, depending upon which is the warp and the weft, and along the bias, which is diagonal to the thread lines. Both materials have their uses and applications, and can be combined in a laminated sail. Conventional Dacron sails differ only from their modern high-modulus Kevlar, carbon and Spectra counterparts in that the materials stretch less for the same weight or to the same degree at a reduced weight.

In a mainsail the highest stresses are in the region of the leech, and therefore high-modulus fibres, with their threads aligned with the stress, are particularly useful in this area. In the forward part of the sail, however, where considerable movement can be introduced by the mast, woven cloth can be used in such a way that its bias stretch compensates evenly for the shape being taken out of the sail. Such composite sails, which combine high-modulus materials with conventional Dacrons, are usually completed with a layer of Mylar film.

HEADSAILS

What is a genoa or jib?

A genoa or jib is a foresail set from a wire luff, a wire headstay or solid headfoil. Genoas overlap the mainsail of a boat by a percentage of from just over 100 per cent to 160 per cent, while jibs do not overlap the mainsail at all.

How does a headsail differ from a mainsail?

A headsail is controlled along only one edge and one corner, and there is less control along that one edge than along the leading edge of a mainsail. Instead of the leading edge being vertical, it is swept back and it usually has no battens. It is the first sail to come into contact with the airflow, and this affects how the helmsman steers the boat.

How should the shape differ from a mainsail?

While the shape of a mainsail is usually only altered in terms of fullness and twist, with the fullness maintained at a fairly constant position, a genoa must be capable of being altered in camber, twist and fullness (draft) position.

Why?

Because the draft position of a genoa affects how the helmsman can sail the boat. If the draft is forward and therefore the entry is blunt the sail will accept a wide range of angles of attack; this may be necessary when the boat is sailing in waves. If the draft is aft the entry will be fine and the sail will only accept a narrow range of angles of attack.

How should a headsail be cambered?

In general the camber should vary between 45 and 50 per cent aft, depending upon the conditions.

In difficult conditions when it is hard to steer the boat accurately – in waves or fluctuating winds – it is useful to drag the camber forward by increasing the halyard tension, since this makes the sail less likely to stall (see above).

In flat-water medium and strong conditions when the boat is easier to steer accurately, the entry of the genoa can be finer since this will enable the boat to point higher. A fine entry also has better lift/drag properties than a blunt entry, but it is more prone to stalling.

As with mainsails, the camber should not be allowed to creep further back than 50 per cent as the wind increases, since this will greatly impair the lift/drag ratio of the sail.

How much camber?

The amount of camber depends upon the wind strength and the sail in question in much the same way as a mainsail. While the position of the camber is altered by the halyard, the amount is altered by the headstay tension – which is controlled by the runner, backstay or sweptback shroud tension. A light genoa might be cambered up to around 15 per cent while a blade jib could be cambered at about eight per cent.

A good way of judging camber and depth is to look up the middle of a sail from the foot, hold a finger horizontally so that it stretches from luff to leech and gauge the shape of the sail against the straight line of the finger. Some sailmakers make calibrated gauges to do the same job but a finger works perfectly well.

▼ You can judge the camber of a sail by looking up at a draft stripe and gauging its shape against a straight line such as the edge of your finger.

How much twist?
Like the mainsail, a genoa must be twisted under certain conditions, and this affects pointing ability. The trim of the genoa affects the size of the slot between the genoa and mainsail, and if the slot is too small for the amount of air passing through the boat will be slowed.

It is easy to see when the slot is too small because the mainsail will backwind excessively. To stop this the genoa must be twisted to open the slot until the mainsail is backwinding only slightly. This is done by moving the car back on longitudinal tracks and upwards on transverse tracks. Twisting the genoa also flattens the base of the sail, depowers the head and reduces the heeling forces.

Twist introduced for this purpose should not be confused with twist used to set the sail correctly for the conditions. A genoa must be twisted when it begins to have a detrimental affect on the mainsail, but it must also be twisted to maximise its own efficiency. Gauging the amount of twist in a genoa is much easier than for a mainsail because, to a large extent, it depends upon the heel angle of the boat.

What is the slot?
The slot is the space between an overlapping genoa and the mainsail. The effect of the slot is to speed the air up by what is known as the venturi effect as the air passes through the narrowing gap. The slot helps to maintain a steady flow around the leeward side of the mainsail which increases the efficiency of the rig – but if the slot is too narrow for the prevailing conditions the effect is to backwind the mainsail too much and slow the boat.

What factors affect the slot?
The size of the slot is affected by the sheeting angle of the genoa, the amount of twist in the genoa, the fullness of the mainsail and the position of the main boom. As the wind increases, there is a tendency to ease the mainsail down the traveller which closes the slot and decreases rig efficiency. So when it is necessary to depower the mainsail in this way it is also necessary to twist off the leech of the genoa to maintain the size of the slot.

How is the sheeting angle set?
The sheeting angle is the angle of a line taken between the tack and the clew of the headsail and the centreline of the boat. In general the closest a headsail can be sheeted to the centreline of the boat is about six degrees, which is what big, heavy close-winded boats like 12-Metres use. In lighter IOR and IMS boats, and dinghies, the average sheeting angle is about 8 degrees, while in Whitbread boats – which don't often sail upwind – the angle is as wide as 10 degrees.

On the water the sheeting angle varies with the type of boat, the wind speed, the overlap of the headsail and the sea conditions. In steady conditions when a headsail is comfortably within its wind range, the sheeting angle can be set at its narrowest for optimum pointing. However, once conditions become unstable – the boat is sailing in waves or variable breezes – or the boat is overpowered, the sheeting angle must be increased to prevent the sail creating too much side force.

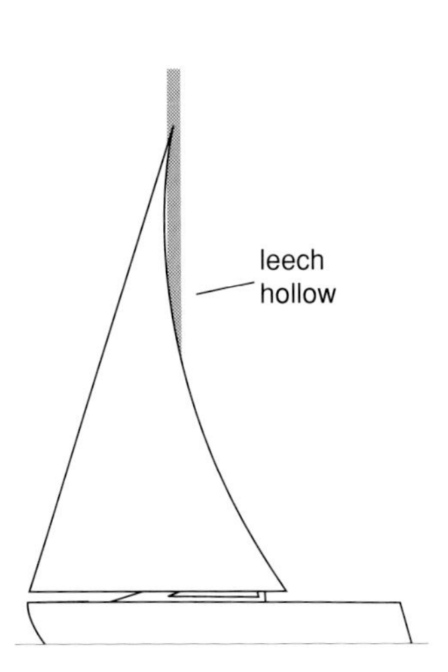

What is leech hollow?

Leech hollow is the amount of concavity needed in a genoa to prevent the leech fluttering. By its very nature a genoa cannot be fitted with battens to support the leech as on a mainsail.

Where should the headsail telltales be sited?

Unlike the mainsail, the critical section of a headsail is its luff – both for the setting of the sail and for the helmsman to steer to – so the genoa telltales are sited near the luff at regular intervals up the sail. The steering telltales are used by both the helmsman and the trimmer: the helmsman needs them to check that the boat is on its most efficient heading, and the trimmer needs them to determine whether the sail should be more sensitive (with a flat entry) or more forgiving (with a blunt entry) or twisted or flattened.

The way to gauge the sensitivity is to bear the boat away from a position where the windward telltale is breaking – when the boat is sailing closer to the wind than the optimum – until the leeward telltale begins to break. If the sail has a flat entry the leeward telltale will start to break almost as soon as the windward one has started streaming. If the sail has a more forgiving round entry, there will be a few degrees through which both telltales are streaming before the leeward one breaks.

The telltales up the luff of the sail are also used to determine the twist in the sail. If all the telltales on the windward side of the sail break together as the boat is luffed up or the sail is eased, then the twist is uniform up the sail. However, it has been found that maximum genoa efficiency occurs when the upper telltales break slightly before the lower ones, with the delay increasing as the waves increase.

How does the material of a genoa compare with that of the mainsail?

In general the panel layout of a headsail is similar to that of a mainsail, with the leech being the most highly loaded area. Therefore directionally-oriented high-modulus materials are used in the leech area with the fibres aligned with the stress up the leech. Towards the luff, the sailcloths are usually used on the bias; this ensures that there is some give in the cloth to allow the shape to be altered by halyard and headstay tension.

In dinghies and keelboats which have just one headsail for all conditions, the cloth must be heavy enough to withstand the highest winds the boat is sailed in. On yachts with several headsails, lighter sails are used for lighter winds since lighter sails cause less pitching, help to lower the overall centre of gravity of the yacht and are easier to set.

SPINNAKERS AND GENNAKERS

What is a spinnaker?

A spinnaker is a three-cornered sail that is set flying from a spinnaker pole and which is attached only at its three corners, rather than along one or more of its sides. A spinnaker is symmetrical about its vertical centreline.

What is a gennaker?

A gennaker is an asymmetric spinnaker which has the characteristics of a spinnaker crossed with a genoa: that is, the air always flows one way across the sail. Like a genoa, a gennaker must always be set with the same corner as the tack and the same corner as the clew. Gennakers are not only asymmetrical in profile – usually with longer luffs than leeches – their section shapes are also more like genoas with the fullness forward rather than in the middle as in a spinnaker.

Why are they shaped as they are?

As downwind sails, spinnakers are designed to maximise downwind sail area efficiently for a wide variety of wind angles. They are required to provide two distinct functions: for reaching they can be regarded as powerful headsails in which the air flows from one edge to the other, while for running they simply catch the air

▶▶ *A symmetrical spinnaker must not be oversheeted, otherwise the acting leech will close down and slow the boat.*

▶ *An asymmetrical gennaker is much more like a genoa, with its draft forward and a clean exit out of the leech.*

K 31
Just in time

like a parachute. To add to the complication, a conventional spinnaker is required to act as an aerofoil when reaching, but since it is symmetrical it must work with the air flowing either way across its surface; this is perhaps the most significant difference between a spinnaker and a mainsail or headsail on which, in general, the air flows only one way across the sail.

What materials and why?

Because spinnakers are used when the wind is on or behind the beam, they operate in lower apparent wind strengths than upwind sails. Therefore the material does not need to be so strong. Also, they are not tacked and are generally not allowed to flog, so they do not wear out as quickly as genoas.

The material used for spinnakers is generally nylon or something similar. Kevlar and Mylar are also used, although it is generally held that some give is desirable in a spinnaker to absorb shock loadings and enable the sail, the luff of which is flying free, to be set easily.

Is size important?

Size can be important but big is not always best. There is little scientific evidence to support the theory that a smaller spinnaker can be more efficient than a larger one, but when reaching, or when a boat is overpowered, a smaller spinnaker is obviously more efficient since it tends to heel the boat less and reduces the tendency to broach. When running, the size is limited by the ability of the

▼ *A spinnaker hoist on the Mumm 36* Jamarella.

sail to stay up: the spinnaker must be filled and supported by the wind before it can drive the boat. For these reasons bigger, heavier sails are sometimes less efficient.

What is the ideal spinnaker shape?

The ideal spinnaker shape depends on whether the sail is to be used for reaching or running. For reaching a flat shape is usually more efficient but the limiting factor is the ability to trim the sail. As with the genoa, a flat entry is more sensitive to angle of attack. Because it is free-flying, this makes a spinnaker twitchy and difficult to set; as soon as the leading edge curls in, the sail collapses. A sail with a rounder entry is more forgiving, and the leading edge will curl over several feet before the sail collapses totally.

For running, projected area is the most important criterion, and a full shape maintains a high projected area more efficiently than a flat shape by supporting bigger shoulders.

In profile the shape very much depends upon the proportions of the rig, but a moderate aspect ratio is important – more so than with the other sails. If it is too short and wide the sail will not be an efficient aerofoil, but if it is too tall and narrow it will tend to 'tube' and create high drag with closed leeches.

Where should telltales be sited?

Since both edges of a symmetrical spinnaker are at some stage the leading edge, the telltales should be placed symmetrically. Telltales on a spinnaker perform the same function as those on a

▲ *Excessive twist in the spinnaker (left) can be prevented by pulling on the twinning line (right).*

headsail – to indicate when the sail is stalling – and should therefore be placed near the leech/luff of the sail on both sides. However, telltales are not critical on a spinnaker since as the sail stalls the leading edge curls inwards, which is much easier to observe. Trimming a spinnaker is simply a matter of keeping the sail on the point of stalling/collapsing.

How does twist affect a spinnaker?

Twist is a useful and necessary feature in a spinnaker. It is controlled by the height of the spinnaker pole and the position of the sheet lead. The height of the pole controls the angle of attack up the height of the spinnaker in the same way that an increase in halyard tension on a genoa rounds up the front of the sail. When the pole is raised the entry is flattened, and when it is lowered the luff is rounded.

With the pole in the correct position the top half of the spinnaker will curl evenly as the sheet is eased. When the pole is too low the luff breaks first at the top. Lowering the pole allows the

▲ *With the pole too far aft (left) the spinnaker is too flat and too close to the boat. If the pole is too far forward (centre) the spinnaker does not project its maximum area. With the pole correct (right) the spinnaker is an efficient shape, yet not shadowed by the main.*

cloth around the tack very little, but moves the upper luff inboard quite a lot, increasing the camber and making it flap. Similarly, when the pole is too high the bottom of the luff breaks first.

The sheet lead position affects the exit of the sail: with the lead aft the base of the spinnaker is flattened compared to the top part, and with the lead forward the leech of the sail is closed and the foot is rounded. When the sheet lead is in the ideal position for reaching the exit of the sail is open without losing too much power out of the head. With the pole in the ideal position this usually means that the clew is slightly higher than the tack.

How should the spinnaker pole be positioned?

This varies for reaching and running. The height of the pole should be set to give the optimum luff shape as described above. Fore and aft, the pole should usually be set as far aft as possible without stretching the foot of the sail around the forestay. In stronger winds and steady conditions, when a flatter sail is required, moving the pole aft helps to flatten the sail. In light conditions or in a seaway when a less sensitive and more powerful sail is required, the pole can be eased forward and the sail rounded up.

How can you get up to a tight reaching mark in light air?

Raising the pole flattens the entry, which enables the boat to point higher. But this also closes the leech, especially if the sail is sheeted hard, which slows you down. You will have to decide whether it is worth doing this, or whether it would be better to drop the kite.

How can you get up to a tight reaching mark in heavy air?

In strong winds the limiting factor is the stability of the boat: even with the main right off the boat will tend to broach as you point higher. Raising the pole, although it flattens the entry, closes the leech and creates more heeling force. So in this case leave the pole and battle it out.

▼ *With the pole too far down the spinnaker starts to collapse; note that the luff curls first above mid-height (left). On the right the pole is too high.*

8 Sail controls

Once the boat is on the water, the job of its designer and builder is over and the task of the crew begins. At this stage tuning is reduced to mast adjustment and sail trimming.

In an ideal world there would be infinite control over every aspect of both the mast and the sails, enabling them to be tailored to suit every conceivable condition. However, the sailor's world is far from ideal, and in many cases systems that are too sophisticated and offer too many control options add both weight and unnecessary, unusable complication. The greater the sophistication, the greater the chance of it being in the wrong trim at any given moment. A simpler boat is often a faster boat, so if in doubt, leave it out!

In race conditions sailing the boat well will always be the prime consideration, and while it might be desirable to adjust spreader deflections or shroud tensions over, say, a five-mile leg, it would almost certainly be disastrous to consider such fine tuning over a 300 or 400-yard sprint. All that said, there can be significant advantages to be had from fitting the right sail controls, calibrating them properly and accurately, and using them judiciously.

What are the various mainsail controls?

The outhaul controls the depth in the lower part of the mainsail. On boats up to about 50 feet, the outhaul is best controlled by a musclebox in the boom which permits easy control without recourse to a winch.

The flattener also controls the depth in the lower part of the mainsail at higher windspeeds, when you need to flatten the bottom of the sail and slightly reduce its area. The flattener is particularly useful because, as it is wound on, it generally raises the boom above the horizontal. This means that as the vang or the mainsheet is wound on and the boom is pulled down, the distance between the corner of the sail and the mast is increased which further flattens the sail. The flattener is usually applied by a winch and locked off on a boom clutch. Raising the clew of the sail also helps to keep the boom out of the water when reaching.

Mainsheet/vang/traveller These three operate in conjunction with one another, and there are many systems for different styles of sailing and different types of boat. There are basically two styles of sheeting the mainsail: conventional sheeting and vang sheeting. Both are intended to control both the angle of the boom to the centreline and the twist in the mainsail leech, but they use different methods with different results.

In a conventional system the vang is slack upwind and leech tension is provided by the mainsheet, while the angle of the boom is controlled by the traveller. The amount of control is limited by the length of the traveller, since once the maximum travel is reached the mainsheet must be eased and this results in a loss of

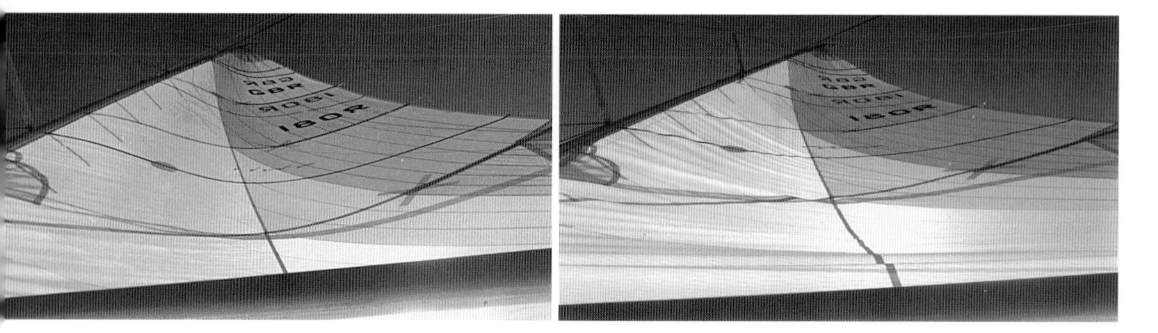

▲ *The main with the outhaul off (left) and with the outhaul on (right).*

leech tension. The further back in the boat the position of the traveller, the narrower the angle of control over the boom.

With a vang sheeting system, the vang controls the leech tension and the mainsheet simply controls the angle of the boom to the centreline. The system is very efficient, particularly on smaller boats, and requires less purchase in the mainsheet which facilitates rapid sheeting in and out.

The disadvantage of the vang sheeting system is that it requires a very stiff boom to enable the vang to exert high leech loads, and in exerting these high loads the vang tension bends the mast forwards at gooseneck level which is not always desirable. In dinghies the vang sheeting system is at its best when gooseneck-level mast rams and A-frames are used to counteract the action of the vang, and in conjunction with a one-to-one mainsheet that trims exceedingly quickly. As boats get bigger the system becomes less practical because high-purchase mainsheet systems are needed, and on large yachts the traveller provides the quickest means of altering the angle of the boom.

The main halyard and cunningham affect the fullness and the draft position of the mainsail. If the mainsail is cut to maximum size it is usual for the halyard to be set to its maximum mark upwind and to control the draft using the cunningham, depending upon wind conditions. The draft in the mainsail should be held at about 50 per cent at all times and, with the halyard set, the cunningham should be used to maintain this. Downwind both should be eased to give the sail maximum fullness, and therefore power, unless the wind is so strong that tension is needed to reduce the power in the sail.

▼ *The main with cunningham off (left) and with cunningham on (right).*

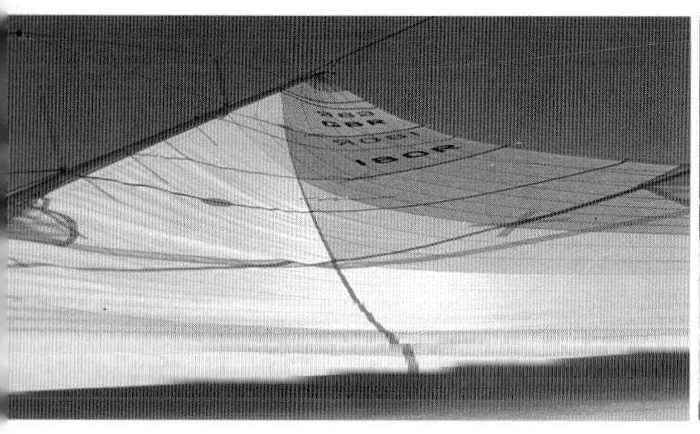

▲ *The effect of too much mainsheet tension (left) and too little (right). In the centre the setting is just right.*

Applying both halyard and cunningham will have a slight flattening effect but they are best regarded as flow position controls rather than fullness controls.

Halyard locks are most effective for upwind sailing and reduce the compression loads in the mast, but they can be prone to jamming and do not allow for easy adjustment of the halyard. The cunningham should be controlled by a purchase system that allows quick and simple operation without recourse to a winch. On some maxi boats the cunningham is operated hydraulically, which has the advantage that it can be controlled remotely by the mainsheet trimmer.

Battens have a profound effect on the shape of a mainsail, particularly the battens of fully-battened mainsails.

The two most important batten characteristics are stiffness and taper. The stiffness of the batten is related to the size of the roach of the sail and the strength required to support the roach. In general, in sails where the battens are not full-length, the battens in the lower part of the sail are stiffer and less tapered than those at the top because they are shorter relative to the chord of the sail. Such battens can be tailored to the sail – so they support the roach without causing a large vertical crease at the inboard end – and usually they need not be adjusted or changed in varying conditions. If the battens are too stiff they will cause a crease and they will need to be shaved down or tapered more.

At the top of the sail the characteristics of the battens are more crucial. Too weak and they will not support the roach, too stiff and a clean shape up top will be impossible to achieve. It is usual to have two or three variations of top batten (or any batten which extends more than two-thirds of the way into the sail). In light airs a thin batten allows the top of the sail to be set at its fullest, while in heavy airs a stiff batten suits a flatter top section and supports the roach. The top battens are usually more tapered too, since they must match a greater proportion of the desired chord shape of the sail.

Full-length battens are more difficult to set up, since not only do they affect the overall shape of the sail, but they also affect the way the sail can be adjusted for different wind conditions. With full-length battens, not only is the stiffness and taper critical, but also the tension with which they are fixed into the sail.

▼ *The stiffness characteristics of full-length battens are critical.*

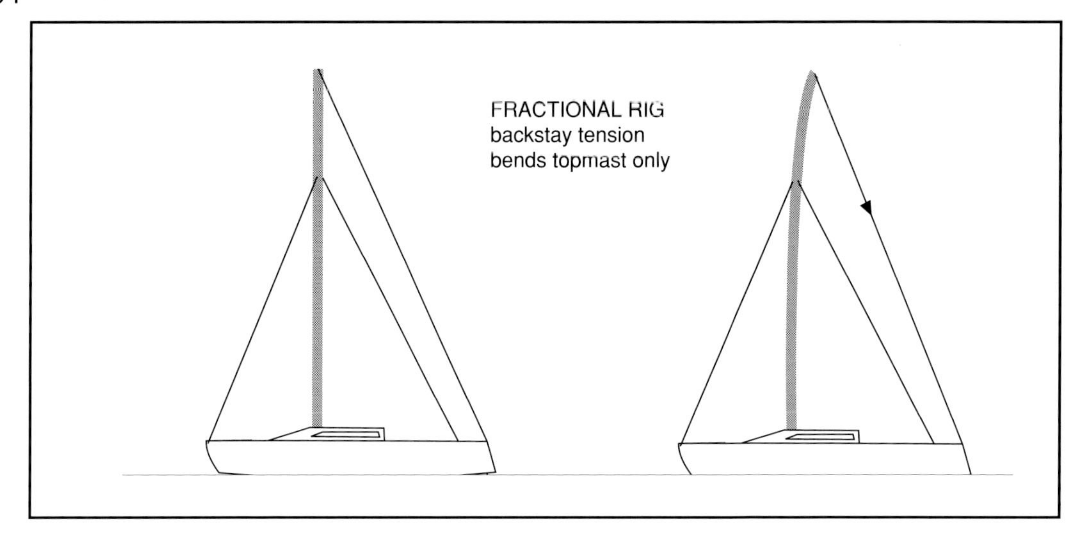

Since full-length battens are generally longer, they are also generally stiffer, with less taper. Because they span the entire chord of the sail, the taper should allow them to bend into the desired chord shape of the sail with the maximum depth in the same percentage position on each batten. However, it is important that they are stiff enough to maintain the sail shape even when maximum leech tension is applied through the vang or mainsheet. Too stiff and they will not allow the sail to take up its designed shape; too soft and they will distort when the sail comes under load, which usually results in the fullness of the sail being dragged forward to the back of the mast while the leech – particularly on large-roached sails – is unable to stand up.

The harder the batten is pushed in, the more camber is induced in the sail. So in light winds the battens are pushed in hard, while in strong winds they are eased until just before the sail starts to wrinkle.

Mast controls
As well as the controls which directly affect the mainsail, there are a multitude of controls and combinations of controls that affect the mast and therefore indirectly affect the mainsail. The effect of each depends upon the nature of the rig – masthead, fractional and so on.

The backstay On a fractional rig with runners this controls the topmast, the fullness of the top section of the mainsail and the openness of the upper leech. For these purposes, where the backstay is never highly loaded, a purchase system that can be operated from both sides of the boat gives the most direct and accurate control.

The backstay on a masthead rig controls the headstay tension and must be more substantial than its fractional counterpart. It is therefore often hydraulically operated. Because the mast of a masthead rig is not so tapered, the backstay has less effect on the

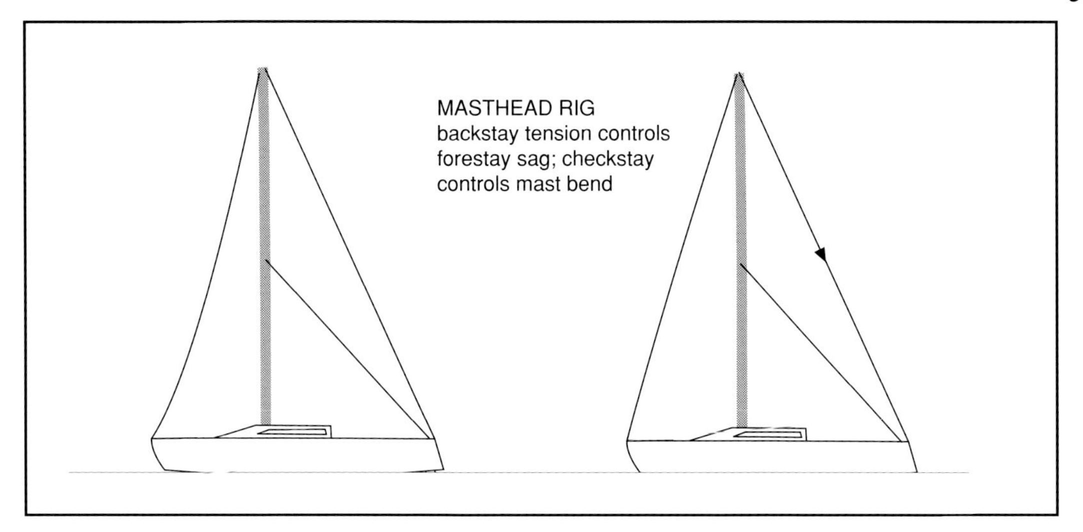

top section of the mainsail. The shape of the upper leech is therefore far less controllable; the leech must be controlled as a whole by the vang or mainsheet.

The runners On a fractional rig the runners are the key to headstay tension. On all but the smallest boats they are operated by a pair of winches and can be calibrated either by a load cell fitted to the forestay or by marks adjacent to the winches. Increased runner tension also increases compression in the mast which, in the absence of checkstays, results in increased bend. On rigs with sweptback spreaders, high runner loads will result in a lack of athwartships rig tension which might allow the mast to bend out of column; this will affect the nature of the slot, twisting the top of the headsail open and reducing its efficiency. So in boats with swept-back spreaders and runners it is important to wind on quite high levels of rig tension in stronger winds to reduce the detrimental effect of high runner loads.

The effect of runner tension – and backstay tension on a mast-head rig – is seen mainly in the headsail where they control the degree of fullness.

▼ *The main with the backstay off (left) and with the backstay on (right).*

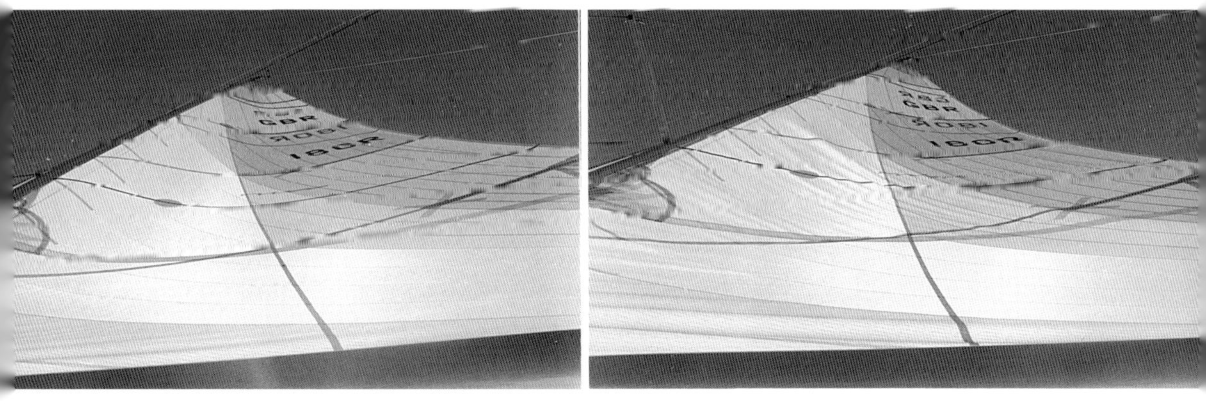

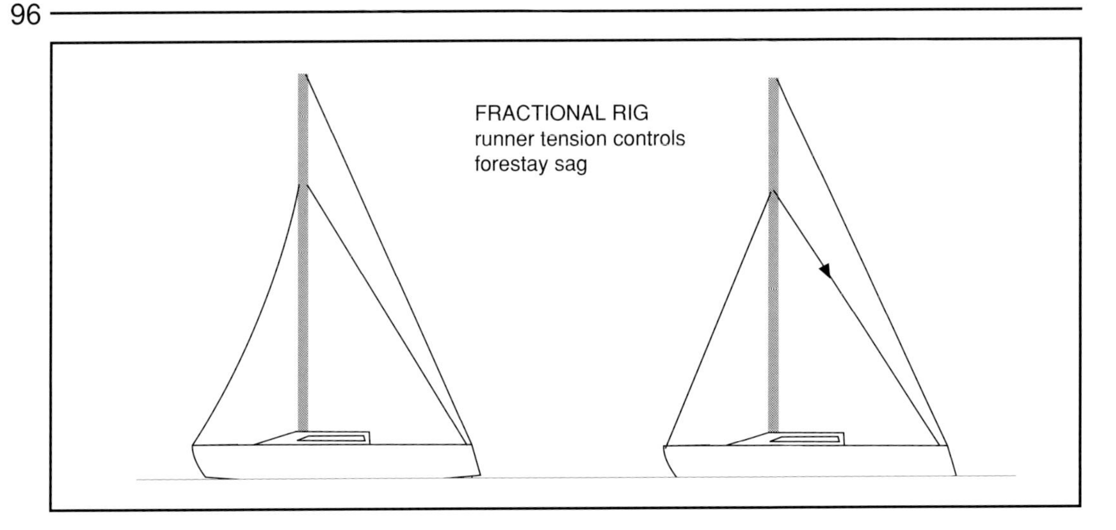

Checkstays are usually found only on fractionally-rigged boats; on masthead boats they have a similar effect but are known as runners. Checkstays control the bend of the mast and thus the fullness of the main body of the mainsail. They operate within certain limits which are governed by the way the mast is set up. These limits are controlled by the amount of pre-bend in the mast, since the checkstays cannot be let out to such an extent that the mast bends forwards at deck level beyond the angle which is predetermined by the heel and the partners (deck level). They also cannot be pulled on past the point where the mast inverts and bends aft above the deck. However, when the mast is set up with the desired pre-bend there is usually enough latitude in checkstay adjustment to fully flatten the mainsail or power it up to its maximum fullness.

The position of the checkstays on the mast is critical to their effectiveness. If they are too high they will not control the base of the mainsail; if they are too low they will work against the inherent pre-bend of the mast and leave the mid-section uncontrolled. So

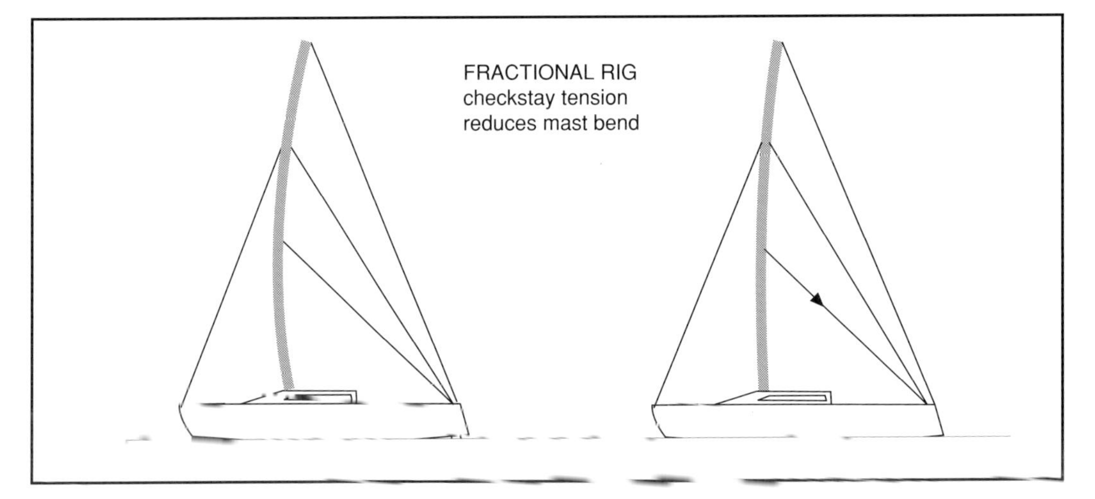

the point of attachment should be at the point of maximum bend of the mast.

The inner forestay is a stay used mainly in masthead rigs and on fractional rigs for long ocean races. It is attached to the mast somewhere on the lower third of its height and is principally fitted to provide stability to the rig in a seaway. However, it can be used to induce some low-down bend when required.

Rams and screws are often fitted to dinghy classes to control pre-bend at deck level, while in keelboat classes chocks are more normal. Being able to move the mast at deck level has a profound effect on the versatility of a mast because it is possible to set up the pre-bend to match the conditions. The deck position is adjusted in conjunction with the heel position depending upon the amount of rake required for upwind sailing.

Spreader adjusters are to be found mainly on dinghies and small keelboats and are used solely on sweptback spreader rigs where the angle of the spreader affects the mast bend. If the shrouds are not deflected fore and aft by the spreader then the spreaders will have no effect on fore and aft bend. However, if the spreaders deflect the shroud aft, then as the load in the shroud is increased – either at the turnbuckle or by an increase in wind – the shroud will

impart a forward force through the spreader to the mast, bending it forward at the spreader root and flattening the mainsail. Similarly, if the spreader deflects the shroud forward, the shroud will in turn straighten the mast and increase the fullness of the sail. The amount the shrouds are deflected by the spreaders depends on the righting moment of the boat and crew, and the cut of the sail.
Adjustable shrouds are again the province of the sweptback spreader rig and are outlawed by some class rules. However, like a mast jack on a boat with in-line spreaders, they increase the rig tension. In a sweptback rig, increasing the rig tension has a varying effect on the mast depending on the spreader deflection described above but, in addition, it affects the tension and hence sag in the forestay. Easing the shrouds will allow the forestay to sag and induce fullness in the headsail, and vice versa. It is therefore usual to sail with less shroud tension in light airs, increasing it as the wind increases. The adjusters can either be powerful purchase systems or compact hydraulic rams.
Diamonds and jumpers were described in detail in the rig section but, to reiterate, they control the topmast of a fractional rig. Diamonds control only sideways bend and jumpers control sideways and fore-and-aft bend. They effectively allow the mast to be of variable stiffness in the top section, and can be adjusted either by turnbuckles at their intersection with the mast or by a hydraulic system or purchase at the base of the mast (if allowed by the class rules). In general, like spreader adjusters and deck-level rams, adjustable diamonds and jumpers fall into the category of complications which can cause more harm than good during a race, particularly a short race. They are often better set up before a race and left alone while the crew concentrates on sailing the boat as well as possible.

What are the various headsail controls?

Most of the controls act on the mainsail, and the genoa and spinnaker are relatively simple by comparison. That is not to say that they are simpler sails to set, just that there are less controls relating to them and therefore the inherent designed shape is more important and less easily manipulated.

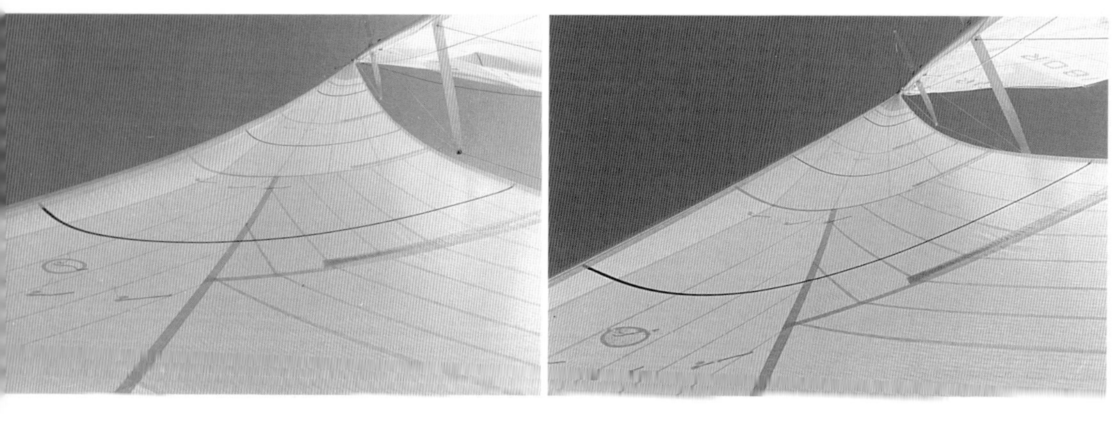

▼ *With the runner off the headstay is slack, giving a full entry (left); winding on the runner gives a tight headstay and a fine entry (right).*

The headstay tension is controlled by the backstay, runners or sweptback shrouds, and affects the fullness of the sail. A tight forestay means a flat sail and vice versa.

The halyard affects the position of maximum fullness and is usually led to a winch or musclebox system. The tighter the halyard, the further forward the fullness, until the wind reaches a level where increased halyard tension only prevents the fullness moving further back rather than bringing it forward. It is at this point, on a boat with several headsails, that the sail is usually changed for a smaller and heavier one.

The cunningham is sometimes used on a genoa for controlling the draft in the lower part of the sail. In truth it is rarely used, although it is useful for changing headsails since the old sail can be set from the cunningham while the tack is released making the change much easier.

The clew position on a genoa is a combination of sheet tension, fore-and-aft position and athwartships position of the sheeting system. The fore-and-aft position affects the relative tensions between the foot and the leech and thus controls twist, while the athwartships position affects the sheeting angle. The narrower the sheeting angle, the higher the boat will point, until the slot between the mainsail and genoa becomes so narrow that the volume of air flowing through is constricted too much and increases the drag. The angle at which that occurs depends upon the conditions and the sail in use, but can usually be spotted by excessive backwinding of the mainsail.

The width of the slot, the sheeting angle and the pointing angle depend also upon the sea state. A boat can point higher in flat water than in choppy water and therefore the sheeting angle must be changed accordingly. For this reason many boats are fitted with athwartships genoa tracks which allow for a high degree of control over sheeting angle.

Similarly, the fore-and-aft position, or the height of the clew, changes with conditions. When the sail is well within its designed wind range and the water is flat, maximum power can be extracted from the sail by maintaining a tight leech and fullness in the base. This is achieved by moving the sheet lead forward and

▼ *With the halyard off (left) the draft is well back in the sail, giving more power but a fine entry. With the halyard too tight (right) the draft is too far forward and the entry is too rounded, killing pointing. In the central picture the halyard and draft are correct.*

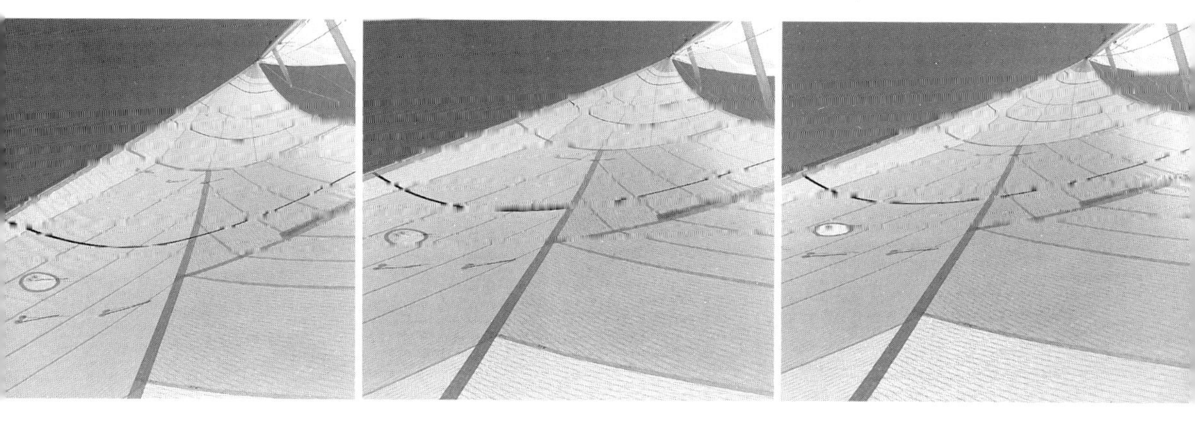

▲ *Moving the car forward gives more power and a tighter leech (left); moving it back gives a flatter sail with an open leech (right).*

effectively bringing the clew of the sail down. As the wind increases the clew is gradually raised which flattens the base of the sail and opens the leech, both of which have depowering effect. In awkward waves, when it is necessary to twist the sail but maintain the power in the base, the sheet lead can be moved forward and the sheet eased, rounding up the base but opening the leech.

Systems for sheeting headsails fall into two categories: longitudinal and transverse. Hybrid combinations of the two have been used with some success, but they are complicated and rarely seen. Longitudinal systems usually consist of two parallel tracks carrying cars that are usually controlled by a purchase system. This gives precise control over leech-to-foot tension but only two sheeting angles. Conversely, transverse tracks give a greater range of sheeting angles but less precise control over clew height; to control the clew height the sheet must pass through a floating

▲ *Excessive backwinding of the main (left) indicates constricted airflow through the slot; this is cured by making the sheeting angle broader (right)*

block which is attached to the athwartships track, meaning that control and calibration of the block is more difficult.

The precision of the sheeting car movement is more significant the higher the aspect ratio of the headsail. Moving the sheeting position of a 150 per cent number one a couple of inches is equivalent to around half an inch on a 100 per cent number three, and therefore the control over the forward headsail cars must be more precise with the smaller sail. On systems which use pins it is usually worth drilling extra holes on the forward tracks or calibrating the forward purchase systems more closely.

What are the various spinnaker controls?

Spinnakers are the most curious of sails, because they need to work with air flowing over them in two directions; both edges are luffs and both are leeches at various times. Their complications

GBR180R

CALIBRATION

The calibration of sailing systems is of vital importance, since it helps with tuning and helps reduce the response time to any oncoming set of conditions. Although no two races are ever the same they are often similar, and setting the leads, halyards and runners to predetermined positions at least provides a starting point for the fine tuning.

In boats where forestay load cells are allowed, these are vital for measuring forestay and thus runner/backstay tension. Failing that the runners or backstay can be marked as they wind onto the winch with a grid on the side deck. The halyards should be marked for each sail, either on the deck where each halyard is marked with its maximum hoist along a grid on the coachroof, or by marking the sail and the forestay at an average setting; if the sail mark is above the forestay mark then the halyard is tighter than average and vice versa. Modern rigging screws are usually manufactured with calibrations, and these should be noted so that the rig can be reset quickly if it has to be taken down and re-erected. The sheeting angles and genoa car positions should also be marked for each sail.

In many ways calibration has a fundamental effect on the way a sailing system is designed, for if it is difficult to regain a fast setting boatspeed will be lost and the benefit of the control system will be lost.

are compounded by the fact that they are set free-flying rather than from a stay, so they are controlled solely from their three corners.

Given that the halyard is usually hoisted to its maximum or pretty near its maximum, the remaining two control points are simply the tack and the clew. The positioning of the sheet blocks is important, but relatively easy to define. In days gone by when masthead rigs proliferated, resulting in large, broad spinnakers, the sheet lead was generally as far aft as possible. Today, with high aspect-ratio fractional rigs, sheet leads have crept forward, and the ideal position is just aft of an angle that bisects the clew when sailing on a reach. This maintains an even leech tension, with the top of the sail slightly open and losing power and the foot slightly flatter than is required for moderate airs. However, since the sheet lead is not usually moved, this is the maximum wind condition; a tweaker is used to bring the sheet down to close the leech for moderate winds.

At the tack the controls are simply the pole height and the amount the pole is wound aft. Pole height affects the fullness of the sail: the higher the pole, the flatter the entry of the sail, while the lower the pole, the fuller the entry. As described in the chapter on sails a flat entry is less forgiving, being harder to trim because, once the leading edge collapses, the whole sail is likely to follow it.

Conversely a round entry can fold quite considerably before a collapse occurs. Therefore in steady conditions the pole is generally set higher than in difficult wind and waves.

Regarding the amount the pole should be brought aft, the further back the pole is brought, the greater the ratio of forward driving force to sideforce and therefore the more efficient the sail. However, the further the pole is brought back the closer the sail will be brought towards the boat – which is not fast. Therefore a rule of thumb is to keep the spinnaker pole roughly at right angles to the wind. Also, the further the pole is brought back the flatter the base of the sail will be. As with raising the pole, this will make the sail more difficult to trim. Therefore a compromise between trimmability and stability must be reached.

The systems available for controlling the pole are many and various, and depend upon the size of the sail, the type of boat and any class rules. Suffice it to say that the pole should be capable of being raised and lowered easily, preferably from the weather side of the boat so that the trimmers can react to course and wind changes quickly. Where possible the pole should always be horizontal to maximise the projection of the sail from the boat.

9 Kinetics

What are kinetics?

Kinetics is the term used for movements designed to enhance or improve performance and speed – such as rocking, rolling, ooching, sculling and pumping. Strictly speaking they relate to movement by the crew but the term can be broadened to encompass styles of sailing and crew positioning as well as sudden actions by the helmsman and trimmers.

Are they legal?

In many cases no – but the rules vary from class to class and regatta to regatta so check them carefully.

How does crew positioning affect a boat's performance?

The weight of a boat's crew can often represent a high proportion of the all-up sailing weight of the boat, and apart from the fact that the crew actually sail the boat, their weight can be used to great affect to control the sailing trim of the boat both statically and dynamically.

Statically, good crew positioning can reduce heel, induce heel, help to prevent broaching and help to reduce wetted surface area. Dynamically, cleverly-timed crew movements can speed up tacking, promote planing and surfing and prevent broaching and loss of control.

How does crew weight affect heel?

It is fairly obvious that the weight of the crew sitting on the weather side of the boat will reduce heel. However, it is worth remembering that placing the heaviest crew members on the widest point of the boat will maximise the available righting moment for a given crew weight.

Crew weight can also be used to induce heel in light airs to get the sails to fill as efficiently as possible.

The times to transfer weight from leeward to windward will vary from boat to boat and will be most easily detected by the helmsman, who can feel the reactions of the boat through the helm. In light airs the boat will be easier to steer with some crew-induced heel, since this will create a small amount of weather helm. However, although the boat may be easier to sail, it may be slightly slower than if sailed upright with neutral helm; less lift is required from the foils in light airs, and if the rudder is not producing lift it creates less drag. So, as soon as the helmsman feels comfortable, the crew weight should be positioned to windward.

When dead running, the boat should be sailed upright or even with a small amount of weather heel if possible. Weather heel puts the centre of effort of the rig above the centre of resistance of the hull and therefore minimises drag since rudder angle is not required to keep the boat sailing straight.

◄ *By 'kiting' this Optimist to windward on the run the helmsman puts the centre of effort over the centre of resistance to minimise drag.*

How does crew weight affect trim?

The effect of crew weight on trim is in many ways more significant and less predetermined. Except in light airs heel should always be minimised, so the crew are usually positioned on the weather side. Trim, however, depends upon the wind, the point of sailing and the waves.

In light airs, placing the crew forward will immerse the narrow sections of the bow and lift the stern out of the water to reduce the wetted surface area. As the wind increases and the boat accelerates towards its hull speed, the crew weight can be used to level the boat out fore and aft to maximise the waterline length.

Sailing upwind in heavy winds, the crew weight positioning depends upon the type of boat. Boats with narrow bows and wide sterns tend to trim down by the bow as they heel, and therefore the crew weight is generally moved aft in an attempt to maintain level trim. Boats with more balanced waterlines and narrower sterns heel more evenly and are not so dependent on crew weight aft for level trim. It is simply a matter of experimentation, but the key, when the boat is powered up, is the cleanness of the wake off the transom.

◄◄ *Wide-sterned yachts with narrow bows tend to trim down by the bow as they heel, so upwind in heavy airs it pays to move crew weight aft*

Upwind in waves it is usual to sail trimmed slightly more bow-down than in flat water, because bow-down trim helps to reduce pitching caused by waves. In dinghies with spinnaker chutes, however, this may not be possible owing to water being shipped through the chute. In general boats with flat bow sections should be sailed bow-down upwind to prevent the boat slamming, since this will ruin the airflow over the rig and the water flow over the keel.

Experiment and experience is also the best way to determine the ideal downwind trim. In light airs the crew are usually moved forward to lift the stern, for the same reasons as before. As the wind increases, the weight is generally moved aft, first to maximise the waterline length and later to improve control as the wind reaches the level where the boat is likely to broach. Moving the weight aft in heavy weather has two effects: firstly it lifts the bow, trims the boat onto its flatter, more stable aft sections and helps to promote planing and surfing. Secondly it moves the centre of lateral resistance aft to help counteract the aft and upward movement of the centre of effort that occurs when the spinnaker is hoisted. Similarly it often pays, particularly in boats with long spinnaker poles that carry lee helm when reaching, to sail with the boat heeled slightly to leeward to balance the helm.

▼ *Trimming aft on the run in strong winds moves the centre of gravity back over the flatter aft sections to promote planing and prevent the bow burying.*

How does crew weight affect pitching?

Since the crew weight is often a significant percentage of the all-up weight of the boat, it can have a significant effect upon pitching. In very light winds, particularly in waves, it is best to concentrate the crew weight at the centre of the boat, so crew that are not needed for sailing the boat are best placed around the top of the keel, either forward and to leeward to help induce heel, or on the centreline and to windward as the wind increases before being called on deck when their full righting moment is needed. In light airs crew have been spotted in some quite ridiculous positions, contorted among the shrouds where they come through the deck or even crammed into the toilet or oilskin locker!

How do crew dynamics affect performance?

Crew dynamics can have a profound effect on performance. However, some of the more useful moves – pumping, ooching and rocking – are outlawed by the rules in many (though not all) cases. During a tack or gybe certain moves are allowed, and in some classes such as Ultras there are *no* illegal crew moves. Yet legal or illegal, it is useful to understand their effects.

Pumping can be compared to pushing off the side of a swimming pool: a swimmer can gain significant impetus from this. In the same way, trimming the mainsail and also the spinnaker (if both sheet and guy are trimmed simultaneously) can give the boat a good push forward. The body of air adjacent to the sail is the spongy equivalent of the side of the pool. As the sail is rapidly trimmed the body of air is jerked aft, inducing an opposing forward force into the sail. In general when racing, pumping is allowed once per wave on the mainsheet and once on the spinnaker guy and sheet.

Ooching is a rapid movement of the crew forward with an abrupt stop. It works because as the crew move forward they gain momentum; when they stop suddenly this momentum is imparted to the boat. Unlike pumping, which is allowed with some restriction, ooching is usually outlawed (although where permitted it can be very effective in marginal surfing conditions for promoting surfing down waves).

In very light airs the most effective crew propulsion method is rocking. The boat is rolled continuously from side to side, often quite violently; the movement must be done in time with the boat's natural roll period and it has the effect of filling the sails, allowing them to generate a driving force, and creating flow over the keel and rudder which enables them to create a driving force too.

What is roll tacking?

Roll tacking is a version of rocking, during which the boat changes tacks. It works rather like rocking, in that the rolling motion increases the apparent windspeed over the sails and causes the foils to generate lift, but it works better in some boats than others. A roll tack or gybe also reduces the amount of helm needed to tack or gybe the boat and therefore reduces the drag significantly.

▲ *Roll tacking a Laser in medium airs. A gentle roll is often more effective than the violent crew kinetics favoured by some sailors.*

The rule on roll tacking states that a boat must not exit the tack faster than it went into it – but it is a difficult one to judge.

To make the most of a roll tack it is necessary to heel the boat to leeward first, then roll to windward as the helm is put over. The amount of helm required will depend on the shape of the boat. The crew should only level the boat out on the new tack once the bow is through the wind and the sails are full.

In dinghies roll tacking is very efficient and it is a highly effective way of making headway upwind in light airs. In keelboats, inducing the heel is more difficult and requires teamwork and timing from the crew. Leaning over the edge of the boat and hanging off the shrouds is highly effective but often illegal, and so any crew dynamics should be developed in conjunction with the IYRU rules and any pertinent class rules.

How fast should a boat be tacked?

This depends on the boat and the conditions, and can only be found through experimentation. Heavy boats which carry their way can be tacked very slowly, with a considerable gain to windward as the boat shoots through the eye of the wind. Lighter boats often need to be tacked more quickly to get the sails drawing on the new tack to accelerate the boat.

In general it is best to put the helm down slowly until the sails are no longer filling, put the boat through the middle of the tack quickly and fall off gently onto the new tack. In stronger winds, the process should be speeded up to level out the boat as the crew weight comes off the rail. The helmsman must then work in conjunction with the headsail trimmer as the boat comes onto the new tack: if the trimmer has the sail sheeted home before the boat is heeled and powered up then the tack is too slow, and if the sail – particularly a large overlapping genoa – has to be sheeted in

using an inordinate amount of winch winding the tack is too fast.

In waves it usual to slow the tack down and to come out of the tack slightly 'wider' than usual; this enables the boat to power up and pick up speed before coming hard onto the wind. It is also essential to pick a flat spot in the wave system; this can often be called to the helmsman by the crew up forward.

In very strong winds, particularly aboard modern light-displacement boats, it is essential to maintain weight on the weather rail for as long as possible and for the crew to move quickly to the new side. Otherwise the boat is likely to heel excessively and slip sideways. The result is not only a loss of windward gauge, but also a stalled keel and difficulty in regaining steerage. For his part, the helmsman must be careful not to lay the boat off too far on the new tack until the boat is moving and the weight is maximised on the new side.

How much rudder should be used to steer the boat?

In general, the rudder should be used as little as possible. Some helmsmen, even successful ones, have a habit of waving the tiller around a great deal but in general the less it is moved the better, because although the rudder provides a turning force it also creates a braking force.

Some dinghy racing schools teach sailors to sail their boats with the rudder removed, which makes helmsmen aware of the effects of sail balance and heel and helps them reduce the amount of helm they use. In general, when the sails are balanced, heeling a boat to windward will make it bear away while heeling it to leeward will make it luff up. This is because the immersed hull shape becomes asymmetrical as the boat heels: the leeward side gets straighter relative to the windward side with heel to windward, and vice versa with heel to leeward.

The fastest sailors use a combination of heel, sail trim and small rudder movements to steer the boat, and use very little helm in flat water. More helm is needed in waves because the boat must be steered not only to the wind, but also to the waves; the relatively violent movements required to steer the boat around a wave does not sap the speed as much as crashing straight through it.

Downwind, violent helm movements often help promote surfing in waves.

What effect do tiller movements have in light airs?

Tiller wagging or sculling is an effective way of driving a boat at quite a pace, particularly when used in conjunction with rocking. However, the rule on sculling states that helm movements not necessary for steering are illegal.

10 Setting up and changing gears

For the purposes of this chapter it is useful to split boats into two types: those with in-line spreaders and those with sweptback spreaders. The difference is fundamental to the way the boat is set up and sailed, because the rig of a boat with sweptback spreaders must be set up before a race and cannot be changed – except in certain dinghy classes where adjustable shrouds are permitted – and the way the rig is set up directly controls the headstay tension. By contrast a rig with in-line spreaders can be adjusted precisely while racing, and is obviously more versatile.

To achieve the sail shapes and mast settings referred to later in this chapter you will have to refer back to the chapter on rigs (Chapter 5), and apply the relevant theory for in-line or sweptback spreaders.

What is the first step?

The first step must be taken before the boat leaves the dock. Any extraneous gear must be removed: extra tools, sails, crew gear and provisions, oilskins, and even cabin doors and tables where allowed. The boat must be sailed in as stripped-out a state as possible within the class rules.

Is this always the case?

Actually no, but for the vast majority of the time it is. If a race can be reliably predicted to be predominantly heavy weather upwind, there is a small advantage to be gained by carrying extra weight to increase the displacement of the boat. This slightly increases the waterline length and the stability, but while marginally increasing the upwind performance, the extra weight will have a detrimental affect on downwind performance so in general it is best to sail as light as possible. In 12-Metres and other heavy boats it often pays to sail upwind with water in the bilge (as long as the water is not *pumped* into the bilge this will not infringe the rules) which can be pumped out off the wind.

How should gear be stowed?

The IYRU and class rules will stipulate the position of many items, but those that are free to be moved should be placed above the keel before the start of the race. This reduces pitching and helps to ensure the centre of gravity of the boat is as low as possible; this increases stability. In light airs the gear should be placed in front of the keel to help trim the bow down slightly and reduce wetted surface area, while in heavy airs the gear can be moved further aft to help maximise the waterline length. Every little helps.

How can conditions be categorised?

Categorising conditions can only be done effectively for an individual boat. The categorisation affects the trimming of sails, the

GBR
GBR
180R
180R

Fractional rig settings, upwind

		very light	light *	medium	strong **	heavy
	true wind (knots)	0–5	4–10	9–18	17–25	25+
GENOA / JIB	headstay/runners (%)	→ 30	30–60	60–100	100	100
	halyard (%)	20	20–50	50–85	85–100	100
	car	normal	forward	norm→ aft	norm→ aft	aft
	sheet tension	slack	medium	med→ hard	hard → eased	eased slightly
MAINSAIL	mast bend	slight	straight	st→ med	max	max
	halyard/cunningham (%)	zero	30	30–70	70–100	100
	outhaul (%)	70	50	50–100	flattener on	flattener or reef
	sheet	eased	medium	hard	100–80%	100–80%
	traveller	up	boom centred	centre → down	down	down 100%
	backstay (%)	zero	0–30	30–70	70–100	100

timing of sail changes, the positioning of the crew and trimming of the boat. However, if they are categorised loosely here, they can be adapted to suit each particular design.

0-5 knots (true wind speed) Very light airs. In light winds, when both the wind speed and the boatspeed are low, the effect of gusts on the apparent wind angle is greater, and the angle fluctuates considerably. It is therefore essential to maintain a rounded entry on the main and genoa to accommodate the variation in angle of attack. This is achieved by tensioning the halyard a little to pull the draft forward.

Unfortunately a rounded entry generally means a full sail, which is not appropriate for these conditions. It is important that the air can flow around the sail without breaking away, and with minimal power in the breeze this means not asking the air to bend too much around a deep aerofoil. It is therefore necessary to keep the sheets eased and the sail twisted with the leeches open to reduce the distortion in the path of the air.

In waves the problem is worse because the pitching of the boat causes the flow of air over the rig to be uneven; it may even be reversed at the top. This increases the variation in apparent wind speed and therefore angle, again particularly at the top of the rig which requires greater twist.

In general, in light airs it is better for the sails to be too open than too closed, and the sails to be undersheeted rather than over-sheeted. In this way the boatspeed will be as high as possible, the only problem being a lack of pointing ability.

In light airs crew weight is usually forward and to leeward, or even down below. The helmsman should use very little helm to steer the boat and steer very slowly to minimise the drag of the rudder.

▲ *You will need to develop a feel for your own boat: with an underpowered boat like a Sigma 38 you will need to change gears later; with an overpowered boat like an Ultra 30 you will need to change gears earlier.*

** When you change to a heavier No 1 genoa, go back to the previous column of settings.*

*** When you change to the No 3, go back to the previous column of settings.*

◀◀ *This yacht is overpowered because the genoa is not twisted off, the mast is too straight and the mainsail is too deep.*

▶ *Insufficient twist in the main gives too much heel (left); with the main twisted and the traveller up the boat comes more level (right). In a gust the traveller can be dumped to depower still further (far right).*

When reaching, the boat will still need to be trimmed down by the bow since it will not be approaching hull speed. Similarly when sailing downwind it will help if the stern is trimmed well clear of the water, and some heel will be advantageous. Even if the course is dead downwind it will be faster to sail very high downwind angles to keep the spinnaker filling, gybing on a series of broad reaches.

4-10 knots In this speed range the boat is in the transitional stage between being underpowered and fully powered up. As the wind increases through this range the leeches of the sails can be hardened to increase pointing ability while maximising their fullness, and therefore power. The sheeting angle of the headsail will be at its minimum in this wind range, and in flat water the draft can be dropped back in the genoa to enable the boat to point as high as possible, narrowing the 'groove' in which the helmsman can steer. In waves more power will be needed as well as a wider groove

for the helmsman, so the draft will need pulling forward using the halyard to give fuller sails.

By the time the wind has reached seven or eight knots true the crew weight will have been transferred to the weather side and positioned so as to maximise the waterline length. In waves it may be necessary to use quite large amounts of helm to steer the boat around the waves rather than smashing into them.

Off the wind the boat should be trimmed for maximum waterline length. As far as sail trim goes, when two-sail reaching the roundness of the base of the genoa versus the tension in the leech is critical. It is tempting to try to balance the upper telltales with the lower ones, keeping the upper leech closed to derive maximum power from the sail, but with a standard genoa, cut for upwind sailing, setting the leech correctly for reaching gives an over-rounded base, which is an inefficient shape for the sail. For this reason Whitbread boats, and others which spend large amounts of time two-sail reaching, usually carry a high-clewed reaching headsail which permits balancing the leech tension with the foot tension without the base of the sail becoming too full.

Again it will be necessary to sail high angles (tack downwind) under spinnaker.

9-18 knots This is the transitional stage when you begin to flatten the sails to depower them rather than improve pointing ability. As the mast is bent to flatten the mainsail, the outhaul is maxed-out and the headstay tension increased the draft will begin to slip aft; this will need countering with increased halyard tension on the genoa and cunningham tension on the main.

At 18 knots, particularly with a large headsail, the main will be beginning to backwind, in which case either the leech of the headsail can be twisted off slightly or the sheeting angle can be increased to open the slot.

The time to take a reef or change a genoa for a smaller one varies from boat to boat, but in general it is better to maintain full mainsail size until the headsail is reduced to a non-overlapping sail, since this maintains the maximum span (hoist) of both sails. The point at which headsails are changed can be judged by the heel angle, the degree of mainsail backwinding and the feel of the boat. Some boats like to be sailed overpowered, particularly those with narrower sterns which are not slowed excessively by heel.

In waves, particularly with a non-overlapping headsail, it may be necessary to increase the sheeting angle or ease the sheet to give the boat the power to get around the waves. Again in these conditions the draft should be kept forward in the sail to give the helmsman a wide steering groove.

At this windspeed the projected area is the most important criterion downwind, particularly when running. The halyard and outhaul on the mainsail should therefore be eased only slightly to maintain the maximum projected area. Similarly, when running under spinnaker the pole should be brought aft as far as possible and the sail flattened and widened.

Downwind VMGs will take the boat much nearer to a dead downwind course, but it will still be best to reach up slightly, depending on the type of boat.

17-25 knots This phase of sail trimming lies just below the point at which little control of the sails is possible. At this stage the mainsail is fully bladed out, and at the point of reefing. The mast is bent to its maximum, the flattener is fully wound on and the backstay on a fractional rig is tensioned hard to flatten the top section of the mainsail and open the upper leech. By this stage the headstay tension should be maximised and probably the halyard as well.

The sheeting angle and sheet tension will vary according to the waves and the backwinding of the mainsail. The sheet should be eased in waves to power up the base of the sail and open the leech to help the boat through the waves, but since the mainsheet or traveller is likely to be eased to the stage where the boom is over the [illegible], it may be necessary to increase the genoa sheeting angle to keep the slot open.

If it is possible to sail dead downwind on the run without broaching spectacularly to leeward, now is the time to do it.
25+ knots By this stage the three corners of each sail will have been separated by the maximum distance – a common maxim for heavy weather sailing. The mainsail should be as flat as possible, with the leech opened to minimise the heeling force. The headstay will be at maximum, along with the halyard, and the leech of the sail will be open to minimise heel and backwinding. The water will rarely be flat, so the headsail sheet will need to be eased to give the sail some fullness, allowing the helmsman to steer around the waves.

In these conditions more than any others it is essential to balance the boat with the sails so that it virtually sails itself. If the headsail is sheeted on too hard the groove will be too narrow; the helmsman will find he is oscillating between sailing too high with the headsail aback and the boat upright, or too deep with the boat heeling over wildly and slipping sideways into the bargain. The right amount of eased headsail sheet, together with the right amount of traveller up or down – depending on the gusts – will require very little from the helmsman in terms of steering except to negotiate the best passage through the waves from the balanced equilibrium position.

When should setup and trim vary from tack to tack?
The fullness of the sails, the position of the fullness and the hardness of the leeches will vary on each tack when the wave train into which the boat is sailing is not perpendicular to the wind. On one tack the boat will be sailing more along the waves, and on the other tack it will be sailing more into them. The tack on which the boat is sailing more along the waves approximates more to flat water than the other, so the boat can be set up with a narrower groove and less twist in the sails, enabling it to point higher. As the boat flips over onto the other tack it will be slamming into the waves and must be set up with a wider steering groove and with greater power from the rig.

Is it possible to tell when to change gears?
The easiest way to tell when a retrim is needed is when the other boats start to pull ahead. By that time, however, distance has been lost, so it is better to anticipate a necessary change. The helmsman is best placed to feel when a boat is overpowered or has become sluggish, but the instruments will also show if the target boatspeeds are not being achieved.

When a boat gets a dose of the 'slows' it is usually caused by oversheeting rather than undersheeting, and by overtight leeches. So if you are going slowly it is usually best to open up the leeches and let the rig breathe. All that will be lost is pointing ability. Conversely, if the boat is sailing fast but pointing badly the leeches are probably too open. If that is not the case then the main traveller may be too far down the track, clogging the slot and failing to provide enough weight in the helm.

11 Instruments

Electronic instruments have become highly sophisticated in recent years, owing to a better understanding of the theories of aero- and hydrodynamics, the computerisation of many functions and the improved accuracy of wind speed and water speed sensors.

Sailing is far from being an exact science, and many of its leading practitioners rely on an innate feel for the motion of a boat through the air and water and have little or no understanding of the scientific principles involved. For these reasons sailing performance instruments tend to be thought of as either indispensable or totally useless. However, as the technology has improved even the diehard seat-of-the-pants sailors are beginning to appreciate the advantages of accurate instrumentation.

What do instruments measure?

The principal measurements – wind speed, wind angle and boatspeed – are measured by sensors at the masthead and under the water. In their most basic form they operate independently of one another to provide apparent wind speed, apparent wind angle and boatspeed through the water. The first level of sophistication is to combine the outputs of the three sensors using a simple computer – often contained within the instruments – to obtain true wind speed and true wind angle (actually modified wind speed and direction as seen earlier).

If an electronic compass is added to the system it is possible to obtain true wind direction (actually modified wind), and finally if the instrument system is patched into a positioning system such as GPS, Decca or Loran it is possible to determine the absolute wind direction and wind speed, and the speed of the yacht over the ground. The latter information is useful for determining the strength and direction of any tidal stream or current.

What is boatspeed?

Boatspeed is the speed of the boat through the water. It takes no account of tidal or current effects.

What is apparent and true wind speed and what are their significance?

The true wind speed is the speed of the wind passing over the land (see Chapter 7). Without detailed tidal information it is difficult to determine the true wind speed and therefore the modified wind speed – as experienced aboard the boat – must be used. This is useful for sail selection on current and future legs since it can be used to estimate the apparent wind that will prevail on any particular leg of a course.

The apparent wind speed is the speed of the wind as altered by the speed of the boat. Upwind the apparent wind speed is higher than the true wind speed, downwind it is lower, while on a beam reach it is the same. The apparent wind speed determines the set and trim of the sails and the upper limits of individual sails.

What is true wind angle and apparent wind angle, and what are their uses?

The true wind angle is the angle of the wind to the land, with the modified wind angle being the angle of the modified wind relative to the boat. The modified wind angle is generally used to detect windshifts, and by either monitoring it visually and making notes or plotting the wind direction against time, it is possible to spot windshift trends and cycles.

Apparent wind angle is the angle of the wind to the boat as experienced aboard the boat. It has a fundamental effect upon sail trimming and steering, since upwind the helmsman steers to the apparent wind and downwind the trimmers trim the sails to the apparent wind. Dead downwind the helmsman and trimmers combine to sail the optimum velocity made good (VMG) which depends on the nature of the boat and the prevailing conditions.

What is the absolute true wind speed and direction?

As we have seen earlier these are the values of wind speed and direction over the land. In non-tidal waters they would be identical to the modified wind speed and modified wind direction, but in tidal waters, particularly in light airs, they may vary considerably.

What is VMG?

VMG, or velocity made good, is a combination of boatspeed and heading. It is of particular interest for upwind sailing and running since it is not possible to sail directly upwind to a destination, and it is often slow to sail directly downwind to a destination.

When sailing upwind, for instance, it may be possible to sail at eight knots and 30 degrees apparent or seven knots and 27 degrees apparent. The highest VMG is the combination of boat-speed and wind angle that will move the boat most rapidly towards its destination, and in this case eight knots at 30 degrees gives a VMG of 6.93 knots whereas seven knots and 27 degrees gives a VMG of 6.24 knots. Similarly downwind, in anything but the strongest wind conditions, sailing a series of broad reaches and gybing downwind will give a higher VMG than running directly towards the mark.

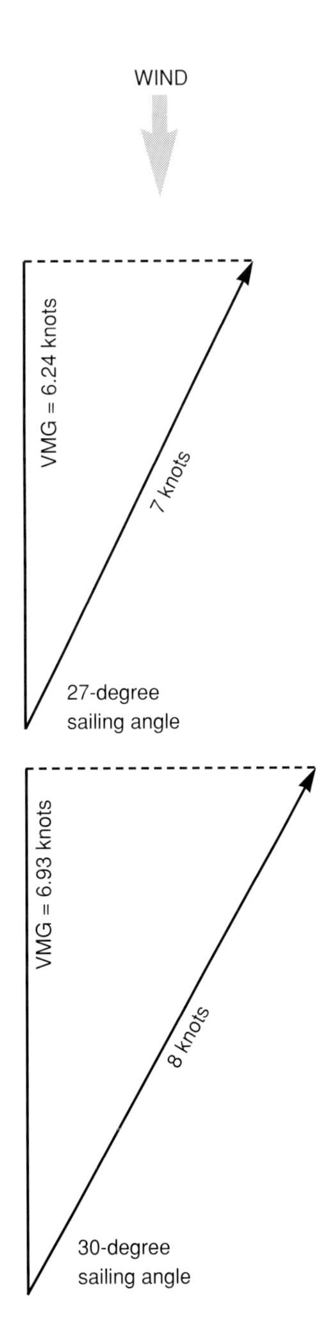

What is a VPP?

A VPP, or velocity prediction program, is a computer estimation of the potential speed of a design. The parameters of the design are input into the computer program which estimates the total drag of the yacht, the driving force its rig develops, the resultant driving force and therefore the potential speed.

What are the shortcomings of VPPs?

Since the computer needs a 'picture' of a boat to estimate potential speed, the accuracy of the prediction depends upon the accuracy of the input information. In practice many assumptions are made, and these compromise the accuracy of the output. Take a mast section, for instance: although it is perfectly possible to define the sectional shape of the mast accurately in the computer, which can then calculate the section drag, it is impossible to account for the drag of every rivet, spreader base or halyard exit. A VPP also takes little account of the effects of sailing the boat in waves. So all along the line there are minute errors in the picture built up by the computer which result in a loss of accuracy.

What use is a VPP?

A VPP is used to produce predicted speeds for the yacht on any point of sailing in any wind speed. The information can be used to handicap the yacht, as with the IMS (International Measurement System), or to produce a polar diagram which is used for tuning by the crew.

What is a polar diagram?

A polar diagram is a radial graph of the speed of a boat relative to its angle to the true wind direction, for a number of different wind speeds. Designers can produce polar diagrams for their designs using velocity prediction programs. These polar diagrams can then be used to optimise the design and, eventually, to get the maximum percentage performance out of the boat. The crew can also collate data from the boat's best on-the-water performance to create their own polar diagram, and this may well be more accurate and useful than a diagram based on a computer-generated VPP.

A polar diagram, which is essentially a graph of a boat's potential, can indicate the closest angle it can sail to the wind, and the optimum angle (for maximum VMG) both upwind and down. It can also show the crossover angle at which a spinnaker becomes more effective than a genoa.

To use a polar diagram the crew simply selects the relevant curve on the polar plot that corresponds to the prevailing wind speed. When sailing upwind the crew is able to read off the maximum theoretical boatspeed they are able to achieve and the angle at which they should sail to achieve it. They then sail the boat to that angle and trim the sails to maximise their speed. The optimum downwind speed and angle can be read off and used in the same way. On a simple reaching leg, the polar diagram will give the speed that should be achievable, and the crew must continually trim the sails to get as near as possible to that speed.

Are polar diagrams accurate?

It is not uncommon for boats to outsail their VPPs: in other words perform faster than the computer predicts they should. In this case the theoretical computer predictions should be modified by actual experience (assuming the instruments are calibrated properly) to create accurate, practical VPPs and polar diagrams. Similarly, when sailing upwind or downwind, experimenting with boatspeed

THE POLAR DIAGRAM

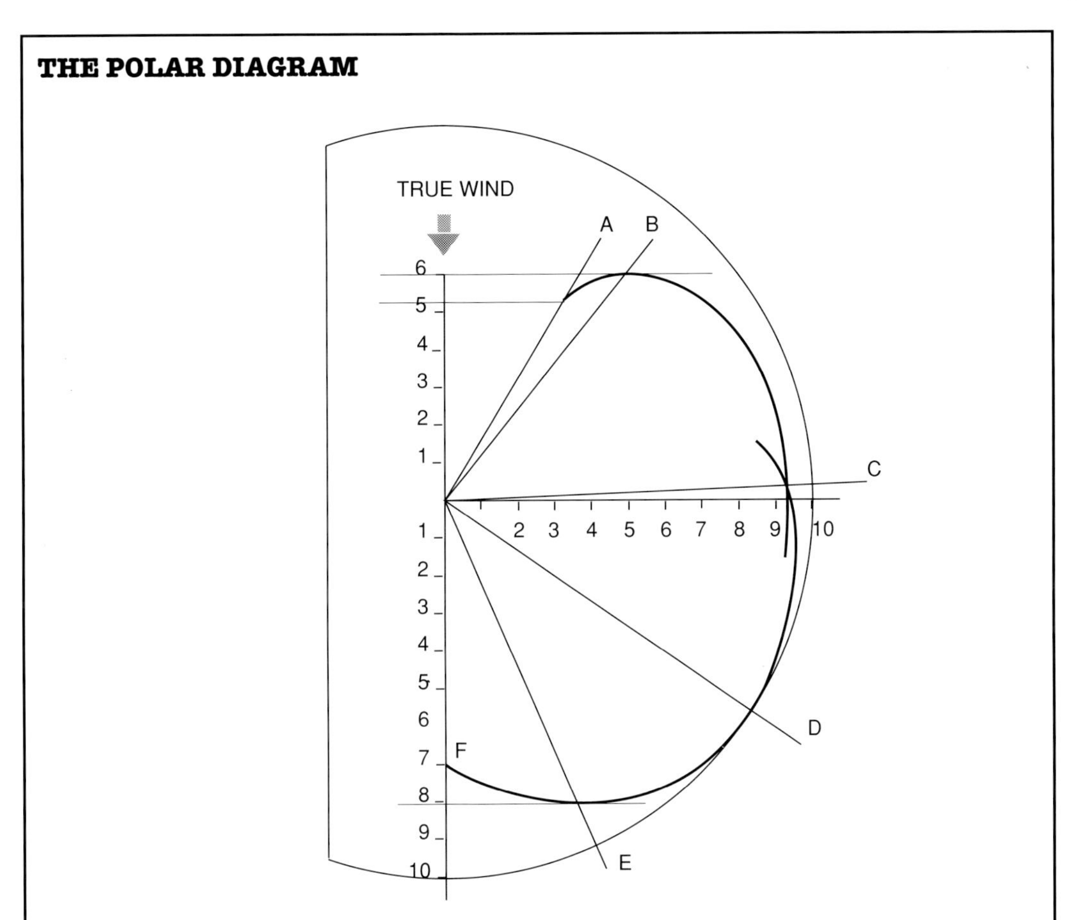

This polar diagram shows the speed of a boat at all headings to the true wind for a constant wind speed. The radius from zero is the boatspeed; the outer circle indicates 10 knots of boatspeed. The angle from vertical is the yacht's heading relative to the true wind, and the boatspeed achieved by the yacht at a range of sailing angles has been plotted to produce a typical 'sliced apple' polar plot.

Angle A is the closest possible angle that the yacht can sail towards the wind, and at this angle the boatspeed is about 6 knots. However, assuming the boat is beating towards an upwind destination its actual speed towards that destination, or VMG, is just over 5 knots (indicated by the horizontal line taken to the vertical axis).

Angle B is the best sailing angle for maximum VMG to windward – some 6 knots – indicated by the horizontal line intersecting with the highest point of the polar curve. Because the yacht is not pinching the boatspeed has increased to almost 8 knots.

Angle C is the crossover point on the reach wher the spinnaker becomes faster than the genoa.

Angle D is the angle at which the yacht achieves its maximum boatspeed in this wind: 10 knots.

Angle E is the best sailing angle for maximum VMG downwind: 8 knots at a boatspeed of 9 knots.

Point F is the boatspeed and VMG the boat achieves running dead downwind.

and angle will show whether the computer has predicted the correct upwind and downwind VMGs. The probable accuracy of a VPP should always be borne in mind when using it. In the first round-the-world race in which Whitbread 60s featured, for example, they all sailed considerably faster than their VPPs.

What is the most useful instrument set-up?

Opinions on what instruments are important. However, our feeling is that the following are desirable:

Boatspeed Used alone, a boatspeed readout is invaluable for showing the effects of any change in trim or sailing angle on speed through the water. The fundamental accuracy of the boatspeed display is not as important as its ability to show trends, unless it is used offshore for dead reckoning navigation.

Wind direction This is used for spotting windshifts and wind direction patterns. Like boatspeed, wind direction is not an absolute reference, particularly in tidal waters where it takes no account of the tidal effect on the apparent, and therefore true wind readings.

True wind speed This is important for verifying boatspeed changes. Only an improvement to boatspeed made in a steady, true wind can be relied upon. Otherwise an assessment of the wind change relative to the boatspeed change must be made. The true wind speed can also be used for sail selection on current and future legs of the course: a quick calculation will reveal the apparent wind speed for any point of sailing and hence the correct sail for any leg.

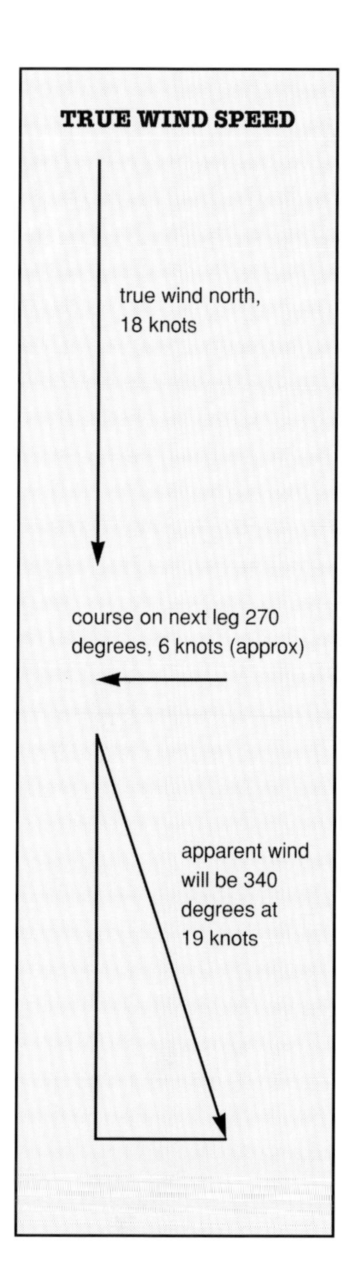

Apparent wind angle Useful for both upwind and downwind sailing, an analogue readout of this information is sometimes preferred to a digital one. Upwind it can be used as a guide in unstable conditions and downwind as a target for a pre-determined optimum VMG.

Compass As well as displaying the heading, the compass can help in spotting windshifts, particularly when close manoeuvring has reduced the accuracy of the electronic instruments.

Obviously there are other readouts such as apparent wind speed, VMG, percentage performance, helm angle, heel angle, water temperature, depth and battery state which can be extracted from modern electronics systems. Some are essential for navigation or pilotage purposes, but others are just distracting or even misleading. The most useful instrument set-up is often the simplest.

What are the limitations of electronic instruments?

Despite being able to give readouts to the nearest one-hundredth of a knot, electronic instruments have several important shortcomings. Boatspeed paddlewheels are affected by weed and turbulence caused by waves, and to counter this some modern systems employ the sonar principle. Masthead instruments suffer from what is known as upwash, which is the flow of air upwards off the sails. This is why most instrument systems are mounted [illegible] forward of the truck of the mast. This location certainly reduces the effects of upwash but does not negate them, and some maxis, former America's Cup 12-Metres and modern America's Cup boats mount two sets of instruments from separate masthead poles – one facing forward, the other aft – to reduce the effects still further.

The biggest shortcoming, however, is the variation in readings from tack to tack. It is not uncommon for wind speed and wind angle readings to vary considerably, which makes tack-to-tack calibration difficult. When the boat is pointing directly into the wind the instruments seldom show this exactly. A similar problem frequently affects the transition from upwind to downwind, and although there is a fine-tuning capability in most instrument systems it is very difficult to eliminate the problem, and it will vary with conditions and sea states. The best remedy is really an awareness of the problems and the tempering of any electronic information with a dose of experience and common sense.

Is it possible to sail to instruments 'blind'?

It is possible to tune into the boatspeed and apparent wind angle upwind and down and to find the numerical 'groove' provided by the instruments. It is done by choosing an optimum upwind boatspeed (from a VPP or from experience) for the prevailing wind speed, and sailing the boat freer to accelerate up to the chosen speed or closer to slow down to it. However, it is essential to monitor the wind speed when doing this because if a decrease in the

wind speed goes undetected the helmsman will find that he is footing off to keep the speed of the boat up – which may not make for optimum VMG at the new wind speed. Similarly, if the wind increases the helmsman will head up.

Off the wind it is a question of choosing a wind angle and sailing to that angle, making sure that a change in the wind direction does not take the boat off course.

These methods can work well in steady and stable conditions. However, it should be remembered that instruments not only give historical information (even if it is only an instant previously) but also average that historical information to update the readout every three or so seconds. So a crew that relies too much on instruments will suffer from reacting to what has happened rather than what is happening.

How can instruments be used for tuning?

Instruments can be vital for tuning, particularly if a designer's polar plot can be input into the system. From a polar plot it is possible to extract target boatspeeds for any conditions and therefore percentage performance figures obtainable by a crew; on any point of sailing it is possible to display how well the boat is being sailed to its polars as a percentage rating. But it should be remembered that neither the polar plot nor the readouts from the instruments are absolute values, and that some experience is necessary to separate the useful information from the garbage and to interpret any electronic results. However, instruments are excellent at displaying trends, for which absolute values are unnecessary.

Are more sophisticated computers of any value?

There are many sophisticated computers available for yacht tuning and racing. The most common systems take inputs from all the yacht's sensors including the electronic compass, plus any positioning system used. Such a system can supply on-screen plots of historical wind data – speed and direction, boatspeed trends and VMG trends. These can be used to detect wind patterns – whether the wind is cyclical around a mean value and, if so, at which stage of the cycle it is at any instant. It can also detect whether shifts are persistent, and moving generally in one direction, or whether the movement is simply random.

Computer systems can also inform a crew of how they have been sailing the boat and how the helmsman has been steering. When linked with a positioning system, the computer can be loaded with the positions of marks, obstructions and navigational waypoints. On inshore races the system can give reasonably accurate estimations of when to tack for marks, corrected for the prevailing current, and can tell the tactician how far to go to a mark and how long it will take. The system will also provide information on course to steer, wind speed and wind angle for future legs. On offshore races the system can predict similar information for marks that are over the horizon or obscured by fog or unseen at night.

Conclusion

Yacht racing technology is improving continuously and the sport of yacht racing is becoming increasingly technical. Yet no matter how sophisticated these systems become, the basic skills of trimming sails, steering boats and making tactical decisions will remain central to success. Fortunately, however highly Bill Koch and others like him rate the science of sailing – and it is important to perfect the science and not underestimate its significance – the art cannot be ignored. In this book we hope we have provided the fundamentals of the science. For the art, well, that's over to you. . . r it is the art that distinguishes the gifted sailor from the purely brilliant one.

For a free full-colour brochure write, phone or email us:

Fernhurst Books, Duke's Path, High Street, Arundel, West Sussex, BN18 9AJ, England

Telephone: 01903 882277 Fax: 01903 882715

Email: sales@fernhurstbooks.co.uk

Or browse our website: www.fernhurstbooks.co.uk